Masa Noguchi (Ed.)

ZEMCH Research Initiatives: Mass Customisation and Sustainability

MDPI

This book is a reprint of the special issue that appeared in the online open access journal *Sustainability* (ISSN 2071-1050) in 2014 and 2015 (available at: http://www.mdpi.com/journal/sustainability/special_issues/customisation-sustainability).

Guest Editor
Masa Noguchi
Faculty of Architecture
Building and Planning
University of Melbourne
757 Swanston Street, VIC 3010
Australia

Editorial Office
MDPI AG
Klybeckstrasse 64
Basel, Switzerland

Publisher
Shu-Kun Lin

Managing Editor
Le Zhang

1. Edition 2015

MDPI • Basel • Beijing • Wuhan

ISBN 978-3-03842-112-2 (PDF)
ISBN 978-3-03842-111-5 (Hbk)

Table of Contents

List of Contributors

Hasim Altan: Sustainable Design of the Built Environment, Faculty of Engineering & IT, British University in Dubai, Dubai 345015, UAE; Welsh School of Architecture, Cardiff University, CF10 3NB Cardiff, UK

María Dolores Andújar-Montoya: Department of Building and Urbanism, Polytechnic University College, University of Alicante, Carretera de San Vicente del Raspeig, s/n, 03690 Alicante, Spain

Ingrid P. S. Araújo: Department of Structural and Construction Engineering, School of Engineering, Federal University of Bahia, Aristides Novis, 2, Federação, Salvador 40210-630, Brazil

Victor Bunster: EDBI Research Group, Faculty of Architecture, Building and Planning, The University of Melbourne, Parkville VIC 3010, Australia

Ali Cheshmehzangi: The University of Nottingham Ningbo China, 199 Taikang East Road, Ningbo 315100, China

Dayana B. Costa: Department of Structural and Construction Engineering, School of Engineering, Federal University of Bahia, Aristides Novis, 2, Federação, Salvador 40210-630, Brazil

Heather Cruickshank: Centre for Sustainable Development, Department of Engineering, University of Cambridge, Cambridge CB2 1PZ, UK

Rita J. B. de Moraes: Department of Structural and Construction Engineering, School of Engineering, Federal University of Bahia, Aristides Novis, 2, Federação, Salvador 40210-630, Brazil

Marcia E. S. Echeveste: Department of Industrial Engineering, Federal University of Rio Grande do Sul (UFRGS), Av. Osvaldo Aranha, 99-5th. Andar, Porto Alegre 90035-190, RS, Brazil

Castiglia Feitosa: Department of Sanitation and Environmental Health of National School of Public Health, University of Sydney, Sydney 2006, Australia

Carlos T. Formoso: Building Innovation Research Unit, Federal University of Rio Grande do Sul (UFRGS), Av. Osvaldo Aranha, 99-3th. Andar, Centro, Porto Alegre 90035-190, RS, Brazil; Building Innovation Research Unit (NORIE), Federal University of Rio Grande do Sul (UFRGS), Av. Osvaldo Aranha, 99, 3° Andar, CEP 90035-190 Porto Alegre, RS, Brasil

FrattariDepartment of Civil, Environmental and Mechanical Engineering, University of Trento, Trento 38123, Italy

Nicola Gasperini: Department of Civil, Environmental and Mechanical Engineering, University of Trento, Trento 38123, Italy

Virgilio Gilart-Iglesias: Department of Computer Science and Technologies, Polytechnic University College, University of Alicante, Carretera de San Vicente del Raspeig, s/n, 03690 Alicante, Spain

Arman Hashemi: Centre for Sustainable Development, Department of Engineering, University of Cambridge, Cambridge CB2 1PZ, UK; Centre for Sustainable Development, Department of Engineering, University of Cambridge, Cambridge CB2 1PZ, UK

Cynthia S. Hentschke: Building Innovation Research Unit, Federal University of Rio Grande do Sul (UFRGS), Av. Osvaldo Aranha, 99-3th. Andar, Centro, Porto Alegre 90035-190, RS, Brazil

Narguess Khatami: School of Civil and Building Engineering, Loughborough University, Loughborough LE11 3TU, UK

Ha-Ryeon Kim: Department of Energy Systems Engineering, Kongju National University, 275 Budae-dong, Cheonan, Chungnam 331-717, Korea

Jin-Hee Kim: Green Energy Technology Research Center, Kongju National University, 275 Budae-dong, Cheonan, Chungnam 331-717, Korea

Jun-Tae Kim: Department of Architectural Engineering, Kongju National University, 275 Budae-dong, Cheonan, Chungnam 331-717, Korea

Diego Marcos-Jorquera: Department of Computer Science and Technologies, Polytechnic University College, University of Alicante, Carretera de San Vicente del Raspeig, s/n, 03690 Alicante, Spain

Andrés Montoyo: Department of Software and Computing Systems, Polytechnic University College, University of Alicante, Carretera de San Vicente del Raspeig, s/n, 03690 Alicante, Spain

Sam Moshaver: School of Architecture, Faculty of Environmental Design, University of Montreal, Montreal, QC H3C 3J7, Canada

Masa Noguchi: EDBI Research Group, Faculty of Architecture, Building and Planning, The University of Melbourne, Parkville VIC 3010, Australia

Cecília G. Rocha: Building Innovation Research Unit, Federal University of Rio Grande do Sul (UFRGS), Av. Osvaldo Aranha, 99-3th. Andar, Centro, Porto Alegre 90035-190, RS, Brazil

Chihiro Shimizu: Faculty of Economics and Business Administration, Reitaku University, 2-1-1 Hikarigaoka, Kashiwa, Chiba 277-8686, Japan

Patrícia Tzortzopoulos: School of Art, Design and Architecture, University of Huddersfield, Queensgate, HD1 3DH Huddersfield, UK

Karl Wagner: School of Business, Programme Master (Eng.) Green and Energy Efficient Buildings, Universiti Kuala Lumpur, Kuala Lumpur 50250, Malaysia

Sara Wilkinson: School of Built Environment, University of Technology, POB 123 Broadway, Ultimo, Sydney 2007, Australia

About the Guest Editor

Masa Noguchi is Associate Professor in Environmental Design at the Faculty of Architecture, Building and Planning, The University of Melbourne, where he also serves as Co-director of the Environmental Design and Building Innovation (EDBI) Research Group. His research is based on design engineering concerning the delivery of zero energy mass custom homes (ZEMCH) and the sustainable communities in developed and developing countries. He has been engaged in various ZEMCH R&D projects globally and in 2007, he designed the *ÉcoTerra® house*, Canada's first net zero energy home built and commercialised through the government's sustainable housing initiative. Dr Noguchi is also known as the co-founder of *ZEMCH Network* and the originator of the mass custom design approach to sustainable affordable housing developments. ZEMCH Network was established in 2010 and today, it consists of 465 partners from over 40 countries. As the coordinator, he has been taking key roles for international collaborations on the initiation and operation of the network's interdisciplinary knowledge transfer events such as the *ZEMCH Mission to Japan*, *ZEMCH International Conference* and *ZEMCH Workshop*.

Preface

In response to growing global warming issues and the constant increase of energy prices, house-builders and housing manufacturers today are becoming more responsive to the delivery of net zero energy and carbon dioxide (CO_2) emission sustainable homes than ever. Within this context, the sustainability may embrace housing economy and adequacy beyond the legitimacy in which the quality barely coincides with individuals' dynamic various needs, desires and expectations. Nevertheless, the industry's business operation tends to follow routines and the close system mode of operation often hinders the enterprises from adopting unfamiliar innovations which may be inevitable in realising the delivery and operation of socially, economically and environmentally sustainable homes.

In this special issue, the notion of "mass customization" was reviewed. This paradoxical concept has been recognised as a means to lessen production costs of end-user products whilst achieving the customisability through economies of scope rather than economies of scale. Housing is no exception. The idea dates back to the 1950s as the gravity became explicit in Walter Gropius' book entitled "Scope of Total Architecture." The essence of mass customisation applied to housing was speculated as he emphasized the need for "standardising and mass-producing not entire houses, but only their component parts which can then be assembled into various types of houses." In fact, housing is a system of energy and environment, composed of a number of parts and components indeed. The choice of the housing design elements need to be made carefully with due consideration of the project's initial and operational cost, quality, and time. Moreover, the location factor cannot be less of a consideration as it encompasses geographical and topographical conditions and local regulations. Location and orientation of housing help secure the optimum use or prevention of sunlight and wind and this affects the building's operational energy consumption and generation which correlate with CO_2 emissions and utility costs.

Mass Customisation is an oxymoron. The term is composed of two opposite notions: mass production and customization. The notion was anticipated in 1970 by Alvin Toffler in his book entitled "Future Shock." In 1987, the term was eventually coined by Stanley M. Davis in his book entitled "Future Perfect." Furthermore, in 1993, Joseph B. Pine II profoundly systematized the general methods of mass-customizing products and services in his book entitled "Mass Customization." In 2009, Frank T. Piller and Mitchell M. Tseng edited a "Handbook of Research in Mass Customization and Personalization" and compiled the R&D activities and outputs delivered by a variety of industries across the globe. The handbook also includes one of the successful mass customization approaches being applied to the housing industry—*i.e.*, modularization of building components. The total number of possible ordered pairs (or combination) of given standard housing components can be quantified. In the approach, the mass customization (MC) has been systematised and visualised simply by making use of a conceptual analogue model as follows: MC = f(PS). In this model, the service sub-system (S) concerns communication platforms that lead the users to participate in customizing their design output while the product sub-system (P) covers production techniques that aim to encourage the standardization of housing components for mass production and dissemination. Standardisation of building components seems to be a limited hindrance to design customisation if communication platforms are well developed. Design-consulting staff and appropriate communication interface are required to facilitate user choice of standard design components. These fundamental design service factors can also be integrated into a comprehensive model: $S = f(l, p, t)$. In this model, the service sub-system (S) is supported by the existence of the location (l), personnel (p) and tool (t) factors and they are necessarily interrelated. Basically, building components can be divided into three categories: volume, exterior and interior. These can be considered the main elements of the product sub-system (P) which can be explained by the following conceptual model: $P = f(v, e, i, o)$. The volume (v) components are used to configure the building's internal space that determines the size and location of each room while the interior (i) and exterior (e) components serve to co-ordinate decorative and functional elements that customize a building. In addition, "o" denotes other optional features such as building amenity and security systems, inclusive design components and renewable energy technologies. In general, fabric and ventilation heat losses are associated with building volume and envelop exposures whilst thermal transmittance links up with materials applied to exterior and interior components.

Most of the net zero-utility-cost housing manufacturers typically in Japan have begun to install a number of renewable energy technologies as standard features rather than options based on their value-added, high cost-performance marketing strategy. The strategy itself is far from new having been applied to a variety of end user products around the world. For instance, although today's automobiles can be produced with lower production costs than those in the past, their selling price does not seem to be affected dramatically by higher productivity. New cars are still generally regarded as expensive; nevertheless, the list of items now offered as standard in new cars, such as air conditioning, a stereo set, airbags, remote-control keys, power steering, power windows and adjustable mirrors, were offered only as expensive options in older models. Clearly, the quality of newer models is much higher than that of older models. The same is true for the housing industry. Quality-oriented production contributes towards the delivery of high cost-performance housing in which high-tech modern conveniences that are installed as options in conventional homes are available as standard equipment (Se). In this context, the product subsystem (P) can further be modified into the following conceptual model: $P = f(v, e, i, o) + Se$. In fact, Japanese housing manufacturers mass-produce net zero-utility-cost customizable homes in which a variety of housing amenities and renewable energy and environmental technologies (e.g., PV, air source heat pump, micro combined heat and power systems, and energy monitoring and control equipment linked to a large-scale lithium-ion battery) tend to be installed as standard features rather than options. Despite the reduction of equipment choices, volumetric, exterior and interior design components still remain substantial options from which the users can choose so as to customize the end product.

In order to deliver a marketable and replicable zero energy/emission mass custom home, or ZEMCH, the strategic balance between the optional and standard features seems to be critical. The optional features may be provided with the aim to enhance design quality (or customizability) that helps contribute to satisfying desires and expectations of individual stakeholders. The standard equipment, on the other hand, needs to be installed in buildings as it aims to exceed product quality whose levels can be adjusted in conjunction with societal demands and requirements. This book encompasses a wide spectrum of hopes and fears around the design, production and marketing approaches to the ZEMCH delivery and operation, and showcases some exemplars budding out in different climates around the globe. It aims to solidify today's diverse expertise in the realm as elicited research paradigms for further exploration and delivery of the sustainable homes that meet the wants and needs of individuals and society.

Dr. Masa Noguchi
Guest Editor

Estimation of Hedonic Single-Family House Price Function Considering Neighborhood Effect Variables

Chihiro Shimizu

Abstract: In the formulation of hedonic models, in addition to locational factors and building structures which affect the house prices, the generation of the omitted variable bias is thought to occur in cases when local environmental variables and the individual characteristics of house buyers are not taken into consideration. However, since it is difficult to obtain local environmental information in a small neighborhood unit and to observe individual characteristics of house buyers, these variables have not been sufficiently considered in previous studies. We demonstrated that non-negligible levels of omitted variable bias are generated if these variables are not considered.

Reprinted from *Sustainability*. Cite as: Shimizu, C. Estimation of Hedonic Single-Family House Price Function Considering Neighborhood Effect Variables. *Sustainability* **2014**, *6*, 2946-2960.

1. Introduction

Economic growth and the progress of urbanization have promoted high-density land use and the construction of high-rise buildings, thereby generating urban problems such as traffic jams, the obstruction of insolation and ventilation, and impaired views. To cope with these problems, cities have actively adopted policies for preserving high-quality urban space through regulating land use and other measures. However, policies such as those on land-use regulations cannot always be achieved with the agreement of citizens due to uncertainties that arise about the effects of such policies. Accordingly, various attempts have been made to measure the effect of land-use regulation and other policies in terms of economic value. The most representative technique for such attempts is a hedonic model that focuses on property values.

However, several problems still remain in the formation of hedonic models [1]. The most significant problem is the identification of estimated variables. Namely, in estimations using a hedonic model, the price is determined as a result of the locational activity (decision-making of a family or a company to buy a property or land) of a standard household (family income) [2]; whereas in an actual city, heterogeneity of family income should be taken as a prerequisite [3]. In this case, data specific to the individual such as income, family types and so on is required. In addition, when a hedonic model is constructed for a wider area, neighborhood variables that can explicitly handle the differences among neighborhoods must be incorporated as explanatory variables. However, in the actual construction of hedonic models, very few studies have appeared in which these variables have been explicitly incorporated [4]. The lack of such studies is due to the problem of bias as a result of omitted variables in the estimation, which occurs because of the impossibility of incorporating all variables that must be considered [1].

When there are unobservable variables such as these, what is the actual extent of the bias that exists in the estimated coefficient? If the extent of the bias were such that it could be ignored, then it would not be necessary to pay that much attention to the problem of omitted variable bias when estimating hedonic functions. The aim of this paper is to clarify the extent of the omitted variable problem that accompanies the existence of unobservable variables.

In terms of methods for addressing the problem of omitted variable bias accompanying unobservable variables, attempts have been made to estimate hedonic functions by performing market segmentation [5,6] or applying a spatial econometric method incorporating spatial autocorrelation in the error term. However, if segmenting regions, while various parameters are estimated in a more region-based form, it is difficult to obtain consistency across adjacent regions due to the segmentation of markets.

Another method is to apply spatial econometrics, which has been developing rapidly in recent years. In terms of methods that consider variables for which spatial observation is not possible, parametric and semi-parametric methods have traditionally been proposed.

As a representative parametric method, an estimation method has been proposed that aims to increase flexibility of fit by means of a high-dimensional polynomial equation using coordinate values (latitude, longitude). This is the so-called Parametric Polynomial Expansion model proposed by Jackson [7], which inputs the squares and cubes of coordinate values and a multi-dimensional cross term into explanatory variables. However, a highly collinear relationship is to be expected between the squares and cubes of coordinate values and the multi-dimensional cross term, and increasing the polynomial equation dimensions in order to increase flexibility of fit leads to a reduction in the degree of freedom. As a result, reduced probability with respect to the degree of confidence of individual estimate values is a concern.

With regard to semi-parametric methods, a Geographically Weighted Regression (GWR) model has been proposed that smooths unobservable geographic differences with coordinate values only, with the purpose of eliminating the effect of those differences on estimated values. Semi-parametric methods are convenient in that they do not assume a function form in advance with respect to the hedonic model. In addition, while it is necessary to introduce multiple geographic attributes to capture geographic local changes with parametric methods, semi-parametric estimation involving local regression performs correction using only coordinate values. However, since complex numerical calculations are performed, the estimation procedure requires an enormous amount of time. Estimation methods that take spatial correlation of the error term into account have also been proposed.

However, all of these approaches do no more than *indirectly* resolve the problem of unobservable variables by devising estimation methods.

Resolving this kind of problem *directly* means addressing it by, insofar as possible, creating observable variables for variables that are considered unobservable through devising GISs and the like. With this in mind, this paper will clarify the extent of the omitted variable problem.

In this study, starting from a hedonic model in which only the building attributes and locational characteristics are considered, we attempt to construct a model that takes into account regional characteristics in a small neighborhood unit and family income attributes, which have not been considered before in Japan. Since such regional characteristic variables and the family income attribute variables affect each other, they show externalities. Therefore, we call them neighborhood effect variables. By comparing a model incorporating these variables with a model that does not, we clarify whether or not omitted variable bias is generated.

2. Hedonic Model with Consideration Given to Neighborhood Effect

2.1. Estimation Model

The following hedonic model was set up using the market of single-family houses and land.

$$\log DP/GA = a_0 + \sum_h a_{1h} X_h + \sum_i a_{2i} Z_i + \sum_j a_{3j} \cdot LD_j + \sum_k a_{4k} \cdot RD_k$$
$$+ \sum_l a_{5l} \cdot TD_l + \sum_m a_{6m} V_m + a_7 u + a_8 v + \varepsilon \quad (1)$$

DP: Price of Detached House Price per Square Meter (Yen)

X_h: Main Variables

- X_1 (GA): Ground Area (Square Meters)
- X_2 (FS): Floor Space (Square Meters)
- X_3 (RW): Front Road Widths (10 cm)
- X_4 (Age): Age of Building
- X_5 (TS): Time to Nearest Station
- X_6 (BD): Bus Dummy
- X_7: $X_5 \times X_6$
- X_8 (TT): Travel Time to Central Business District

Z_i: Other Variables

- Z_1 (RT): Market Reservation Time (week)
- Z_2 (SD): South-facing Dummy
- Z_3 (LaD): Land Dummy

LD_j: Location (Ward) Dummy (j = 0…J)

RD_k: Railway Dummy (k = 0…K)

TD_l: Time Dummy (l = 0…L)

V_m: Neighborhood Effect Variables

- V_1 (FR): Floor Area Ratio/FAR
- V_2 (ZD): Zoning Dummy

$V_{3\text{–}6}$: LU_n/Land Use Condition

- V_3 (LU_1): *Average Building Area*
- V_4 (LU_2): *Standard Deviation of Building Area*
- V_5 (LU_3): *Building Density*
- V_6 (LU_4): *Building Density*^{2}

$V_{7\text{–}8}$: HC_o/Household Characteristics

- V_7 (HC_1): *Rate of Office Worker*
- V_8 (HC_2): *Average Building Area per Household*
- V_9 (EX) Noise

u, v: Longitude, Latitude[3]

In the model and in accordance with previous studies, the following variables were set as major explanatory variables (*Xh*) that determine house prices: ground area (*GA*), floor space (*FS*), front road width (*RW*), age of building (*Age*), time to nearest station (*TS*), bus dummy (*BD*), cross term of *TS* and *BD* (*TS* × *BD*), and travel time to the CBD (*TT*) (see, for example, Shimizu and Nishimura, Diewert and Shimizu [8–10]). In addition, some dummy variables were considered including a location (ward) dummy (*LDj*) and a railway dummy (*RDk*) as locational characteristics as well as a time dummy (*TDl*) to take into account the difference in the transaction time, on top of the variables for the time elapsed between putting a house on the market to the conclusion of the contract (market reservation time, *Z1*), the direction the windows face (*Z2*), and a dummy variable indicating that a transaction is made only for land (land dummy, *Z3*). The model composed of only these basic variables is referred to as Model-1.

In addition to these variables, the model was expanded to one that takes the neighborhood effect into account, which is the major aim of this study. First, to handle the neighborhood effect explicitly, variables (*Vm*) were incorporated that represent the neighborhood effect [11].

Vm mainly comprises the following: zoning restrictions imposed by urban planning, which is publicly available information, the local land-use conditions (*LUn*), which can only be obtained by actual on-site observation, the residential household characteristics (*HCo*), and environmental externality (*EX*).

First, as variables related to land-use regulations, the following two variables were set that are easily obtained from publications and advertisements on the Internet: V1: floor area ratio (*FR*) and V2: zoning dummy (*ZD*). The model in which these variables, which represent generally observable residential characteristics, are incorporated is referred to as Model-2.

Furthermore, the following variables, which can only be obtained by direct observation of properties: V3 to V6: local land-use conditions (*LUn*), and V7 to V8: household characteristics (*HCo*), were incorporated.

Even when the model is controlled by the incorporation of such neighborhood effect variables, it is still predicted that the unobserved variables problem remains. To deal with this, geographical coordinate data (longitude and latitude) based on Jackson's method [5] was incorporated.

The model in which the neighborhood effect variables (*Vm*) and the coordinate data (u, v) are incorporated is referred to as Model-3.

2.2. Data

2.2.1. Single-Family House Price

The analysis data used in this study is summarized in the following data for transactions concluded in the 23 wards of Tokyo (621 km^2) within one year, from January to December 2000.

The prices of single-family houses and land in "Residential Information Weekly" published by RECRUIT Co., as the main information source, were used. This weekly magazine provides information on the characteristics and asking price of properties and land and includes the historical price data for individual properties, from the time the properties are first listed in the magazine until they are removed because of sale or other reasons. The prices are refreshed on a weekly basis. Three points of price information are available in the magazine: (1) the initial asking price (first offer price) when the house was placed on the market; (2) the price upon removal from the magazine (estimated purchase price: first bid price); and (3) the actual transaction price, which is collected as a sample for statistical purposes. The first asking price represents the seller's desired price rather than the market value. In contrast, the transaction prices are enough to estimate the hedonic model considering neighborhood effects.

From the information published in Residential Information Weekly, it was decided to use the price at the time of removal from the magazine as a result of contract conclusion, as the explained variable of the model. The price at the time of removal from the magazine is the first price offered by a prospective buyer; such a bid is offered through a process in which several particulars of the characteristics and price are disclosed to the market via the magazine, and the price is decreased until the buyer responds to that information in a manner that is opposite to that of an auction. Thus, this price indicates the upper range of the market price but it is extremely near the transaction price. (Comparing results of 962 samples we observed between (2) last listed price and (3) transaction price, more than 95% of the samples were equal).

2.2.2. Data Regarding House Characteristics

The price of a single-family house is determined by information on the land and the building. Ground area (*GA*), floor space (*FS*), and front road width (*RW*) were adopted as the numerical data representing the attributes of the land and building. The age of the building is the period from the construction of the building to the conclusion of the transaction. To take into account whether the house's windows are south-facing or not, a south-facing dummy (*SD*) was defined. In addition, when a transaction is only for land without a building, it is handled by a land dummy (*LaD*).

Furthermore, the convenience of public transportation from each house location is represented by the travel time to the "Central Business District" or the CBD (*TT*) and the time to nearest station (*TS*). The travel time to the CBD is measured in the following way. First, the CBD was defined. The metropolitan area of Tokyo is composed of 23 wards with Tokyo as its center containing a dense railway network.

Terminal stations were designated as the center of major business districts. The terminal stations chosen included six on the Yamanote Line: Tokyo, Shinagawa, Shibuya, Shinjuku, Ikebukuro, and Ueno, as well as Otemachi as the central station of the Tokyo Metro (Teito Rapid Transit Authority). Then, the average travel times during the day from each station to the seven terminal stations were investigated, and the minimum value as the travel time to the CBD for each property were set.

Regarding the time to the nearest station, times for different means of transportation are available. There are three means of transportation: by foot, by bus, and by car. However, the data for analysis only include houses within walking distance or bus-transportation distance. Therefore, any difference in the distance between the former and latter is controlled by a dummy variable (bus dummy: *BD*). In addition, the walking time (in minutes) is recorded when the house/land is within walking distance, and walking distance from the house/land to the bus stop and the riding time from the bus stop to the nearest station (in minutes) are recorded for houses/land in a bus-transportation area. The time to the nearest station (*TS*) is defined as (walking time to nearest station) + (walking time to bus stop) +

(riding time from bus stop to nearest station). Then, for a bus-transportation area, the cross term of the constant dummy variable with the time to the nearest station (*TS*) is incorporated in the bus dummy.

These variables are factors attributed to single-family houses; regional price differences are expected to exist as well. Therefore, a location (ward) dummy (*LD*) was set to reflect differences in the quality of public services and differences in area prestige. Furthermore, since most of the residential ground developments in the Tokyo metropolitan area have been carried out along railway lines, the price structure of houses differs along each railway line; therefore, a railway line dummy (*RD*) was defined. Moreover, a time dummy (*TD*) was used to control differences in temporal price changes.

The sale price of each property is also affected by the fluidity and depth of the market. The time taken until the contract is concluded is considered to be affected by the period and location and by the transaction activity in the market. These kinds of market factors are explained using variables such as market reservation time (*RT*). The RT is the time between the date when a land/house is placed on the market by a seller and the date when the transaction is concluded. Properties with a long market reservation time are regarded as having a price higher than the equilibrium price or participating in a thin market. Conversely, properties with a short market reservation time are regarded as being in a market with high fluidity or having a price close to or lower than the equilibrium price. For the purpose of this study, the market reservation time was defined as the time between the first listing of the property in the magazine and its removal from the magazine as a result of contract conclusion.

2.2.3. Data on Neighborhood Effect Factor

In defining variables based on local land-use conditions (LUn), individual building data was used from the “Survey on Land Use in Tokyo” in fiscal 2001. In this survey, data on 1,662,088 buildings are arranged as geographic information system (GIS) data (polygon data) presenting the conditions of use, area, structure, and other parameters of buildings. With these data, (1) the number of buildings; (2) the average area of all buildings; and (3) the standard deviation of building area in each square of side length 500 m, which corresponds to the tallying region of the national census, were calculated. The average area serves as a variable with the same characteristic as the building density, and the standard deviation of the building area was assumed as a proxy variable representing the appearance of the town. It is assumed that, in a region with a small standard deviation, houses have similar sizes along well-ordered streets, whereas in a region with a large standard deviation, the town’s appearance is not well-controlled.

As household characteristic (HCo) data, the following items were set as variables in accordance with the national census: the ratio of households with an office worker (specialist and engineer employees + management-level employees + clerical employees/total households). The ratio of households with an office worker is assumed as a proxy variable representing the differences in the income and academic background in a region. This can be done because these employees are generally known to have a more academic background and higher income level on average than households with other occupations.

Data on the floor space per household was also incorporated. The average area calculated from the "Survey on Land Use in Tokyo" is the average area of buildings existing in a mesh regardless of ownership, whereas the average area observed in the national census means the floor space attributed to a household, which is different from the former area index. It is considered that, while the former is a proxy variable representing ambient environment, the latter can be used as a proxy variable representing assets or income.

Table 1 shows a list of the explanatory variables. The observation data consist of 13,822 transactions concluded between January and December 2000.

Table 1. List of explanatory variables.

Symbols	Variables	Contents	Unit
GA	Ground Area/square meters	Ground Area	m^2
FS	Floor space/square meters	Floor space	m^2
RW	Front Road Widths	Front Road Widths	10 cm
AGE	Number of years since construction	Period between the date when the data is deleted from the magazine and the date of construction of the building.	Years
TS	Time to nearest station	Time distance to the nearest station (Time by Walk or Bus)	Minutes
BD	Bus dummy	Whether the time distance includes riding time of bus 1 Not including bus time 0	(0,1)
TT	Travel Time to central business district	Minimum railway riding time in daytime to seven terminal stations in 2000	Minutes
RT	Market reservation time	Period between the date when the data appear in the magazine for the first time and the date of being deleted.	Weeks
SD	*South-facing dummy*	Windows facing south 1 other directions 0	*(0,1)*
LaD	Land dummy	Transaction is only for land without building 1 otherwise 0	(0,1)
$LD_j\ (j = 0, \ldots, J)$	Location (Ward) dummy	*j th administrative district* 1 other district 0	(0,1)
$RD_k\ (k = 0, \ldots, K)$	Railway line dummy	*k th railway line* 1 other railway line 0	(0,1)
$TD_l\ (l = 0, \ldots, L)$	Time dummy (quarterly)	*l th quarter* 1 other quarter 0	(0,1)
FR	Floor Area Ratio	Floor area ratio	%

Table 1. *Cont.*

Symbols	Variables	Contents	Unit
ZD	Zoning dummy	Land use zoning i 1 other land use 0.	(0,1)
LUn (n = 1,2,3,4)	Land use condition	(1) Average building area, (2) Standard deviation of building area, (3) Building density, and (4) Building density squared	
HCo (o = 1,2)	Household Characteristics	(1) Rate of office worker and (2) Average building area per household	
EX	Road traffic noise	Road traffic noise	db
u, v	Longitude, Latitude	Geographical coordinates	

2.2.4. Statistical Distribution of Single-Family House Prices

Table 2 shows descriptive statistics of the major variables.

Table 2. Summary statistics of single-family house data.

Variables	Average	Standard Deviation	Minimum	Maximum
DP: Price of Detached house price (10,000 Yen)	7211.86	4384.60	1050.00	39,800.00
GA: Ground Area (m^2)	114.99	75.07	10.11	797.35
DP/GA	68.67	24.08	12.00	392.00
FS: Floor space (m^2)	76.83	69.37	0.00	649.08
RW: Front Road Widths	4.85	2.48	2.00	50.00
Age: Age of Building(years)	5.28	8.75	0.00	41.33
TS: Time to the nearest station (minutes)	9.64	4.36	0.00	36.00
TT: Travel Time to Central Business District (minutes)	11.45	6.08	1.00	33.00
RT: Reservation Time (week)	2.39	2.53	0.00	35.00
01/2000–12/2000	n = 12,954			

The average single-family house price is 72.11 million yen, the minimum value is 10.50 million yen, and the maximum value is 398.00 million yen, with a fairly large standard deviation of 43.84 million yen. The data include a wide range of single-family houses from small dwellings to the over-100-million-yen large ones. The average unit price is approximately 0.68 million yen/m^2 with a small variation.

The minimum ground area is 10.11 m^2 and the maximum area is 797 m^2, showing a large variation. The minimum floor space is 0 m^2 for land-only transactions and the maximum value is 649 m^2, with an average of 77 m^2.

Regarding the age of the buildings, while the average age is 5.28 years due to the inclusion of a large number of newly built houses, the maximum value is 41.33 years, with a right-skewed distribution.

Regarding the time to the nearest station, there are properties with a minimum value of 0 min that are located in front of a station; the maximum value is 36 min, with an average value of 10 min. On average, while many properties are conveniently located, some are located in areas with inconvenient public transport.

The minimum travel time to the CBD is 1 min, which indicates that there are properties located adjacent to main terminal stations. The maximum travel time is 33 min, and the average time is 11 min. The variation is small because the subject of the analysis is the 23 wards of Tokyo, wherein a dense railway network has been developed.

3. Estimated Results

3.1. Formulation of Hedonic Models

Model-1 was formulated on the basis of Equation (1) using the market for single-family houses and land in the 23 wards of Tokyo. The results are shown in the first column of Table 3.

Since the adjusted R-square value is 0.644, the formulated model has a fairly high explanatory power as a primitive model. The model was basically linear, and the selection of variables constituting the zoning dummy and railway line dummy was performed by a best-subset selection procedure based on Malow's CP when the major variables (*Xh*) were incorporated.

Table 3. Estimated results of modified hedonic models.

Independent Variables	**Model-1**		**Model-2**		**Model-3**	
***X*:Property Characteristics**	**Coefficient**	***p*-value**	**Coefficient**	***p*-value**	**Coefficient**	***p*-value**
Constant	4.7131	***	4.6550	***	−53.42	***
GA: Ground Area	−0.0014	***	−0.0014	***	−0.0015	***
FS: Floor space	0.0008	***	0.0008	***	0.0007	***
RW: Front Road Widths	0.0169	***	0.0157	***	0.0143	***
Age: Age of Building	−0.0115	***	−0.0115	***	−0.0116	***
TS: Time to the nearest station	−0.0105	***	−0.0099	***	−0.0100	***
BD: Bus Dummy	−0.1773	***	−0.1691	***	−0.1622	***
TS × *BD*	0.0028		0.0024		0.0029	
TT: Travel Time to CBD	−0.0128	***	−0.0118	***	−0.0100	***
***Z*:Other Property Characteristics**	**Coefficient**		**Coefficient**		**Coefficient**	
RT: Market Reservation Time	−0.0006		−0.0008		−0.0007	
SD: South Face Dummy	0.0292	***	0.0285	***	0.0301	***
LaD: Land Dummy	−0.2824	***	−0.2836	***	−0.2932	***
Location (Ward) Dummy ***LDj (j = 0,…,J)***	Yes		Yes		Yes	
Railway/Subway Line Dummy ***RDk** (k = 0,…,K)*	Yes		Yes		Yes	
Time Dummy ***TDl (l = 0,…,L)***	Yes		Yes		Yes	

Table 3. *Cont.*

Independent Variables	**Model-1**		**Model-2**		**Model-3**	
***V1-2*: Neighborhood Effects**	**Coefficient**		**Coefficient**		**Coefficient**	
FR: Floor area ratio	–	–	0.0002	***	0.0003	***
ZD: Zoning Dummy (Commercial use)	–	–	−0.0056		−0.0148	*
***V3-8*:Neighborhood Effects by GIS**	**Coefficient**		**Coefficient**		**Coefficient**	
LU1: Average building area	–	–	–	–	0.0018	***
LU2: Standard deviation of building area	–	–	–	–	−0.0001	***
LU3: BLD: Building Density	–	–	–	–	0.0001	***
LU4: BLD^2	–	–	–	–	0.0000	***
HC1: Rate of Office worker	–	–	–	–	0.0058	***
HC2: Average building area per household	–	–	–	–	0.0029	***
EX: Road traffic noise	–	–	–	–	−0.0160	***
latitude	–	–	–	–	0.4028	***
longitude[3]	–	–	–	–	0.0000	
Number of observations =	12,954		12,954		12,954	
Adjusted R-Square =	0.644		0.645		0.662	
AIC =	−3327		−3372		−4000	

Notes: Method of Estimation: OLS(Ordinary Least Squares regression); Dependent Variable: *DP*: Price of Detached House (in log) per m^2; Signs ***, **, and * represent the estimated parameters are significant at 1%, 5% and 10% respectively.

Using Model-1 as a starting model, other models were formulated wherein consideration was given to neighborhood effect variables (Model-2 and Model-3 in Table 3).

First, in Model-2, the urban-plan zoning dummies for house use, commercial use, and industrial use, as well as the floor area ratio were added. After incorporating the variables adopted in Model-1, variables were selected to be added by the best-subset selection procedure. As a result, the commercial-use zoning dummy and the floor area ratio were adopted.

Next, Model-3 was formulated by incorporating the ambient land-use conditions (*LU*), and the household characteristics (*HC*) obtained in the national census, and the latitude and longitude coordinates. After incorporating the variables adopted in Model-2, the cubed term of the latitude and the linear term of longitude coordinates were included and variables corresponding to V3 to V8 were added.

Here, with the expansion of the estimation model, we verified the correlation coefficient between the key variables Ground Area (*GA*), Floor Space (*FS*), Front Road Width (*RW*), Age of Building (*Age*), Time to the Nearest Station (*TS*), and Travel Time to CBD (*TT*) and the expanded variables for Model 2 and Model 3; the results showed that between the six key variables and the added variables, there were none which had a significant correlation coefficient. As a result, the problem of multicollinearity does not occur. In addition, the comparison here was made using a simple linear model, rather than assuming a non-linear model as in Shimizu, Nishimura, and Karato [12].

3.2. *Effect of Neighborhood on House Price*

3.2.1. Ambient Land-Use Conditions: LU

According to Model-3 regarding ambient land-use conditions (*LU*), the estimated average building area per mesh (500 m × 500 m) had a positive sign and was statistically significant at the 1% significance level, indicating that an increase in the average building area positively affects the single-family house price. In contrast, the estimated standard deviation of the building area had a negative sign and was statistically significant. In addition, while the estimated building density had a positive sign, the estimated second-order term had a negative sign. A simple computation shows that as the building density increases, the house price increases, until the 95% of the building density value in the sample, at which point the house price decreases. In a region with a large average building area and uniform areas of buildings, the local environment is pleasant and the appearance of the town is well ordered, which causes house prices to increase. However, as the variation in building area increases, the local environment deteriorates, resulting in a decrease in house price.

Regarding building density, *i.e.*, the degree of concentration of buildings in a regional unit, the primary term, which affects property values linearly, has a positive effect, whereas the estimated second-order term had a negative effect and was statistically significant. This suggests that an appropriate degree of building concentration has a positive effect on property value, whereas high concentration has a negative effect.

3.2.2. Household Characteristics: HC

Regarding the household characteristics surveyed in the national census, the office worker ratio, as defined in Section 2.2.2, and the average building area per household were adopted in Model-3. Both of these variables were estimated to have positive signs and were statistically significant at the 1% significance level. Since the academic background and income of people working in these occupations are expected to be high, the office worker ratio variable is considered to serve as a proxy variable representing both the ability to buy a house and the academic background. The average building area per household was adopted independently of the average building area (GIS data), and similarly it was estimated to be a positive externality.

In houses in a region with a large average building area according to GIS data, the likelihood that attention is paid to the façade (appearance) of the houses, such as tidy hedges, is high, so that an externality due to such physical contributions to the urban space is expected; this was considered to be the reason why the estimated average building area has a positive sign and is statistically significant. Meanwhile, at the level of the household unit, a large building area means a large area owned by a household; accordingly, the variable of the average building area shows characteristics of a proxy variable representing the assets and economic power of households. Thus, these two effects are estimated independently.

In other words, this suggests that in a region of households with high income and considerable assets, house prices tend to be high. Such a result is consistent with an economic model proposed by Rosen [2], thus indicating that house prices are not determined only by physical factors such as the houses themselves.

3.2.3. Other Factors

The coordinate values (*u*, *v*), the cubed term of latitude and the linear term of longitude, were adopted. This feature suggests that, in this model, geographical attributes remain that cannot be absorbed by the incorporation of travel time to the CBD, the land (ward) dummy, the railway line dummy, and the neighborhood effect factors.

3.3. Omitted Variable Bias

Through the above analyses, we have shown that neighborhood effect variables significantly affect house prices. Consequently, if we do not take these variables into consideration, omitted variable bias is generated in the values estimated by hedonic models. Therefore, a comparison was made of the estimated parameters of main variables in the three models formulated in this study, *i.e.*, Model-1, Model-2, and Model-3 (Table 4), and the bias levels were measured. The presence of omitted variable bias is confirmed by comparing the estimated parameters in base models to the confidence intervals of estimated parameters in modified models.

Table 4. Comparison of the estimated parameters among three models.

Estimated Parameters	Model-1	Model-2	Test *	Model-2	Model-3	Test *
GA: Ground Area	−0.00143	−0.00139		−0.00139	−0.00147	b
FS: Floor space	0.00077	0.00077		0.00077	0.00070	
RW: Front Road Widths	0.01693	0.01575		0.01575	0.01432	c
Age: Age of Building	−0.01146	−0.01148		−0.01148	−0.01162	
TS: Time to the nearest station	−0.01052	−0.00994		−0.00994	−0.01002	
BD: Bus Dummy	−0.17730	−0.16906		−0.16906	−0.16220	
TT: Travel Time to CBD	−0.01279	−0.01177	b	−0.01177	−0.01003	a
Neighborhood Effects	No	Yes		Yes	Yes	
Neighborhood Effects by GIS	No	No		No	Yes	
Unobserved Spatial Effects	No	No		No	Yes	

* The test verified whether or not the estimate values of right column are within the 99%, 95%, and 90% confidence intervals for the parameters estimated for left column; “a”, “b”, “c” indicate the estimated parameters in the left column are not in the 99%, 95%, and 90% confidence intervals, respectively, of the estimated parameters in the right column.

3.3.1. Comparison between Model-1 and Model-2

When Model-1 and Model-2 are compared, we see that the “travel time to the CBD” is estimated lower in absolute value in Model-2, but no difference is apparent in the estimated parameters of the other major variables between the two models.

Specifically, in Table 4, we verified whether or not the Model-2 estimate values are within the 99%, 95%, and 90% confidence intervals for the parameters estimated for Model-1. The results show that only "Travel Time to CBD" is not within the 95% confidence interval.

The difference between these two models is caused by the presence or absence of the land-use regulation dummy and the floor area ratio. Since the floor area ratio and land-use regulation cause a change in the effectively available area, the occurrence of large differences between the statistics estimated using the models with and without these variables is expected. However, a clear difference is confirmed only in the estimated parameter of "travel time to the CBD" and not in other variables such as "ground area" and "front road width", which are thought to correlate with neighborhood effect variables such as "floor area ratio" and "land use regulations".

This result does not necessarily indicate the absence of the problem of biases in the estimated parameters induced by omitting neighborhood effect variables.

Thus we continue to incorporate more specific neighborhood effect variables measured by GIS (Model-3), rather than just incorporating those variables easily obtained from publication such as "floor area ratio" and "land use regulations" (Model-2).

3.3.2. Comparison between Model-2 and Model-3

Next, Model-2 and Model-3 were compared. In Model-3 the variables LU, HC, and NOi are added and the coordinate data are incorporated. No clear changes were noted in the estimated values of building characteristics such as "floor space", "age of building", and locational characteristics such as "time to the nearest station" and "bus dummy" which are thought to correlate with housing characteristics. However a certain bias was observed for locational characteristic variables such as "ground area" and "front road width", as well as "travel time to the CBD".

In Table 4, in a similar manner to the comparison above, we verified the level of probability that the Model-3 estimation parameters would not fall within the confidence intervals for the Model-2 estimation parameters. This showed that the estimated parameters did not fall within the 90% confidence interval for "Ground Area", the 90% confidence interval for "Front Road Width", and the 99% confidence interval for "Travel Time to CBD". In other words, these show the probability that there will be a difference for the respective estimated values.

This means that the effect of each variable considered as a neighborhood effect variable in Model-3 was integrated in the major variables in Model-2.

First, regarding "front road width (*RW*)", in residential districts in the 23 wards of Tokyo where houses are densely built, it is predicted that, as RW increases, insolation and ventilation are improved and public spaces are widened, leading to an improvement in the local environment. Therefore, when neighborhood effects are not considered, it is possible that such environmental factors were absorbed in RW in the estimation.

Regarding the variables representing the convenience of public transport such as "travel time to the CBD", when environmental factors and household characteristics were not considered, the house price was estimated to decrease with the physical distance from the CBD. However, as the distance from stations with large commercial areas (a large CBD) increases, environmental factors such as building density improve. When this is taken into account, as the distance from the center of a city or from regions with a large amount of commerce increases, these environmental factors may cause the house price to increase. Thus, when the neighborhood effect and household characteristics are taken into consideration, the effect of "travel time to the CBD" is less negative.

As described, we can see that without taking into consideration the neighborhood effect variables, non-negligible omitted variable bias is generated in the hedonic model of the housing market in the 23 wards of Tokyo.

4. Conclusions

In this study, to estimate hedonic price functions, a focus was placed on the problem of omitted variable bias and an attempt was made to clarify the effect of incorporating neighborhood effect variables as a means of reducing this bias. In concrete terms, starting from a hedonic model wherein only the building attributes and locational characteristics are considered, the model was expanded by incorporating neighborhood effect variables and observed changes in the estimated parameters.

By considering the neighborhood effect variables represented by the environmental variables and the incomes of households in a small neighborhood unit, the objective was to expand the initial hedonic model and to clarify the effect on reducing omitted variable bias. As demonstrated by the series of estimated results, when the neighborhood effect variables were not considered in the estimation using hedonic models, bias was generated in the estimated parameters.

The application of hedonic models as an evaluation method for the effects of Green buildings has been actively attempted in Japan. We hope that this study contributes to the evaluation of such policies.

Acknowledgments

I'm grateful for helpful comments and suggestions from Erwin Diewert, Kiyohiko Nishimura, and Tsutomu Watanabe. I also thank Yasuhito Kawamura for support preparing the dataset.

Conflicts of Interest

The author declares no conflict of interest.

References

1. Ekeland, I.; Heckman, J.J.; Nesheim, L. Identification and Estimation of Hedonic Models. *J. Polit. Econ.* **2004**, *112*, 60–109.
2. Rosen, S. Hedonic Prices and Implicit Markets, Product Differentiation in Pure Competition. *J. Polit. Econ.* **1974**, *82*, 34–55.
3. Kanemoto, Y.; Nakamura, R.A New Approach to the Estimation of Structural Equations in Hedonic Models. *J. Urban Econ.* **1986**, *19*, 218–233.
4. Asami, Y.; Gao, X.L. Effect of Environmental Factors on Housing Prices: Application of GIS to Urban-Policy Analysis. In *GIS-Based Studies in the Humanities and Social Sciences*, Okabe, A., Ed.; Taylor & Francis: Boca Raton, FL, USA, 2006; pp. 211–228.

5. Goodman, A.C.; Thibodeau, T.G. Dwelling Age Heteroskedasticity in Repeat Sales House Price Equations. *Real Estate Econ.* **1998**, *26*, 151–171.
6. Goodman, A.C.; Thibodeau, T.G. Housing market segmentation and hedonic prediction accuracy. *J. Hous. Econ.* **2003**, *12*, 181–201.
7. Jackson, J. Intraurban Variation in the Price of Housing. *J. Urban Econ.* **1979**, *6*, 464–479.
8. Shimizu, C.; Nishimura, K.G. Biases in Appraisal Land Price Information: The Case of Japan. *J. Prop. Investig. Finance* **2006**, *26*, 150–175.
9. Shimizu, C.; Nishimura, K.G. Pricing structure in Tokyo metropolitan land markets and its structural changes: Pre-bubble, bubble, and post-bubble periods. *J. Real Estate Finance Econ.* **2007**, *35*, 475–496.
10. Diewert, E.; Shimizu, C. Residential Property Price Indexes for Tokyo. Available online: http://www.economics.ubc.ca/files/2014/04/pdf_paper_erwin-diewert-13-7-residential-property.pdf (accessed on 28 April 2014)
11. Shimizu, C.; Nishimura, K.G.; Asami, Y. Search and Vacancy Costs in the Tokyo Housing Market: An Attempt to Measure Social Costs of Imperfect Information. *Rev. Urban Reg. Dev. Stud.* **2004**, *16*, 210–230.
12. Shimizu, C.; Nishimura, K.G.; Karato, K. Nonlinearity of Housing Price Structure—The Secondhand Condominium Market in Tokyo Metropolitan Area. Available online: http://www.csis.u-tokyo.ac.jp/dp/86r1.pdf (accessed on 1 May 2010).

Identification and Characterization of Particulate Matter Concentrations at Construction Jobsites

Ingrid P. S. Araújo, Dayana B. Costa and Rita J. B. de Moraes

Abstract: The identification and characterization of particulate matter (PM) concentrations from construction site activities pose major challenges due to the diverse characteristics related to different aspects, such as concentration, particle size and particle composition. Moreover, the characterization of particulate matter is influenced by meteorological conditions, including temperature, humidity, rainfall and wind speed. This paper is part of a broader investigation that aims to develop a methodology for assessing the environmental impacts caused by the PM emissions that arise from construction activities. The objective of this paper is to identify and characterize the PM emissions on a construction site with different aerodynamic diameters (PM2.5, PM10, total suspended particulates (TSP)), based on an exploratory study. Initially, a protocol was developed to standardize the construction site selection criteria, laboratory procedures, field sample collection and laboratory analysis. This protocol was applied on a multifamily residential building construction site during three different construction phases (earthworks, superstructure and finishings) aimed at measuring and monitoring PM concentrations arising from construction activities. The particulate matter was characterized in different particle sizes. Results showed that the higher TSP emissions arising from construction activities provoked environmental impacts. Some limitations to the results were identified, especially with regards the need for a detailed investigation about the influence of different construction phases on PM emissions. The findings provided significant knowledge about various situations, serving as a basis for improving the existing methodology for particulate material collection on construction sites and the development of future studies on the specific construction site phases.

Reprinted from *Sustainability*. Cite as: Araújo, I.; Costa, D.; de Moraes, R. Identification and Characterization of Particulate Matter Concentrations at Construction Jobsites. *Sustainability* **2014**, *6*, 7666-7688.

1. Introduction

Environmental protection is an important issue throughout the world. Building construction and operations have large direct and indirect effects on the environment [1,2]. Pollution sources resulting from construction processes include harmful gases, noise, dust, solid and liquid wastes [3].

Despite developments, the construction industry is still considered an important source of atmospheric pollution due to particulate matter emissions, causing negative impacts on human health and the environment. Worldwide, it is estimated that air pollution caused by PM2.5 (particulate matter less than 2.5 μm in diameter) in the atmosphere is responsible for approximately 0.8 million premature deaths and 6.4 million years of life lost annually [4].

Dust and PM emissions can come from a number of sources during construction activities, as well as from off-site vehicles associated with the construction works and on-site machinery (off-road emissions), including both static and non-road mobile machinery [5,6]. The detrimental health impacts of PM emissions are not confined to the construction site, since fine particles (particulate matter smaller than 2.5 μm in diameter) can travel further than coarser dust (particulate matter between 2.5 μm and 10 μm in diameter) and can therefore affect the health of people living and working in the surrounding area of the site [5,6].

Emissions of particles and dust from construction can also have an impact on indoor air quality in the neighboring area [7]. Dust and other air pollution from demolition and construction can impact greatly on the health and quality of the lives of people working on and living close to these sites if they are badly managed [7].

The growth of construction and its environmental impacts highlight the importance of the need for sustainable construction processes and ways to manage sustainability measures. It is very important to predict the environmental impacts of a construction site and how they can be prevented before starting an activity [8].

The environmental pollution theme has been well studied in the academic area, and general pollution standards and methodologies exist. In addition, the literature review makes reference to some studies concerning the measurement of particulate matter on construction jobsites from an occupational health workers' perspective [9–11]. Despite knowledge of this, the development of a specific methodology to measure and evaluate particulate matter on construction sites from an environmental perspective remains relatively unexplored. Therefore, there is a need to measure the relevance of emission sources of particulate matter during the different construction phases, as well as to identify mitigating measures to reduce its impact, generating better air quality in the workplace.

Additionally, the Brazilian code for air pollution does not establish quality standard levels that consider the generating source and the chemical composition of the material, considering only the mass concentration, and it mentions the potential toxic risk. Therefore, there is a need for the establishment of standards on the particulate matter emission level arising from construction sites [6,12].

This paper is part of a broader investigation that aims to validate a methodology for assessing the environmental impacts caused by PM emissions arising from construction activities and aims to determine air pollution concentrations caused by particle emissions during the different construction phases. This assessment aims to contribute to the identification and evaluation of appropriate measures of mitigating impacts caused by particulate matter. The objective of this paper is to identify and characterize the environmental impacts caused by PM emanating from construction sites with different aerodynamic diameters (PM2.5, PM10, total suspended particulates (TSP)) based on an exploratory study. This study is part of the Technologies for Sustainable Construction on Social Housing Jobsites Project (Cantechis Project), sponsored by FINEP (Financiadora de Estudos e Projetos) (Brazilian Innovation Agency) and developed by four Brazilian universities: the Federal University of São Carlos (UFSCar), Federal University of Bahia (UFBA), Federal University of Rio Grande do Sul (UFRGS) and University of São Paulo (USP).

2. Background

2.1. Atmospheric Pollution and Particulate Matter

According to Daly and Zanetti [13], air pollution is defined as the anthropogenic emission of harmful chemicals that alter the chemical composition of the natural atmosphere and have an adverse effect on the health of living things, an adverse effect on anthropogenic or natural non-living structures or reduce the air's visibility.

An air pollutant is any substance emitted into the air from an anthropogenic, biogenic or geogenic source that is either not part of the natural atmosphere or that is present in higher concentrations than in the natural atmosphere and that may cause short-term or long-term adverse effects [13].

Most dust particles are too big to be inhaled, but can cause eye, nose and throat irritation and lead to deposition on cars, windows and property [7]. They include evidence of cardiovascular disease exacerbation as noted by multi-center studies assessing the association between ambient air pollution and cardiovascular disease mortality [14]. Other recent studies are providing incomplete, but intriguing results suggesting that particle-induced pulmonary and systemic inflammation, accelerated atherosclerosis and altered cardiac autonomic function may be part of the pathophysiological pathways linking particulate air pollution with cardiovascular mortality [15].

Particulate matter (PM) is composed of inert carbonaceous cores with multiple layers of various adsorbed molecules, including metals, organic pollutants, acid salts and biological elements, such as endotoxins, allergens and pollen fragments [16]. PM is classified in the following types.

- "Total suspended particulates" (TSP) is a name given to particles of sizes up to about 50 μm. The larger particles in this class are too big to get past our noses or throats, and so, they cannot enter our lungs. They are often from wind-blown dust and may cause soiling of buildings and clothes. However, TSP samples may also contain the small PM10 and PM2.5 particles that may enter into our lungs [17,18].

Total suspended particulates (TSP) with additional subcategories of particles smaller than 10 μm (PM10) and particles smaller than 2.5 μm (PM2.5) are discussed. PM can exist in solid or liquid form and includes smoke, dust, aerosols, metallic oxides and pollen. Sources of PM2.5 include all types of combustion, such as motor vehicles, power plants, residential wood burning, forest fires, agricultural burning and some industrial processes [17,18]. Size and chemical composition are among the most important parameters influencing the way in which airborne particles interact with the environment. Deposition from the atmosphere onto soil, vegetation and other surfaces is a function of particle size [19].

The extent to which airborne particles penetrate the human respiratory system is determined mainly by size, with possible health effects resulting from the presence of toxic substances [19]. Visibility degradation is known to be a function of both the size and composition of the airborne particles [19]. A clear distinction is that particles smaller than 2.5 μm penetrate into the alveoli and terminal bronchioles; larger particles of up to 10 μm will deposit primarily in the primary bronchi, and much larger particles (up to 100 μm) will deposit in the nasopharynx [20].

By far the greatest number of particles fall into the ultrafine size range, consisting of PM with a diameter of 0.1 μm or less (PM0.1). These ultrafine particles (UFPs) dominate the surface area of particulate pollution, but do not contribute largely to the PM mass [17]. These size fractions arise primarily from combustion emissions and secondly from particles produced by gas-to-particle conversion processes. They are inherently unstable and grow into larger particles through coagulation and condensation [20]. These particles are dominated by sulfates, nitrates, organic carbon (OC) and, namely, elemental carbon (EC). Ultrafine particles (UFPs) present a particular health threat in that their small size allows greater lung penetration and onward passage across the air-blood barrier [20].

The lifetime of PM2.5 in the atmosphere is typically several days in the absence of precipitation, because fine particles have a negligible sedimentation rate and are not removed rapidly by dry deposition processes [21]. They can be transported thousands of km and remain in the atmosphere for a number of days. Coarse dust particles can settle rapidly from the atmosphere (within hours) and normally travel only short distances [17,18].

The chemical composition of particulate matter is an essential element of information for assessing its source and health effects. Knowledge on chemical composition allows the identification of the potential harmful effect of PM [22].

2.2. Air Quality Standards

Standard references are established in order to differentiate between a polluted and non-polluted atmosphere. These are determined by air quality standards, which define the maximum concentration levels of an atmospheric pollutant, which considers that higher concentrations will affect the health and safety of the population (primary standard), as well as result in damages to the flora, fauna, material and environment as a whole (secondary standard) [23]. These standards were established based on scientific studies concerning the effects produced by a specific pollutant.

The Brazilian National Environment Council (CONAMA) Resolution Number 3 published on Aug, 1990 [24] states that when area classes are not established, primary standards should be adopted. In Brazil, the standardized pollutants are TSP, smoke, sulfur dioxide (SO_2), inhalable particles, carbon monoxide (CO), ozone (O_3) and nitrogen dioxide, as shown in Table 1. The CONAMA Resolution Number 3 also states the critical air pollution scenarios, which are attention, alert and urgency (see Table 2).

Table 1. Brazilian National Air Quality Standards according to Company of Technology Environmental Sanitation—CETESB (2004). TSP, total suspended particulates; PM, particulate matter.

Pollutant	Averaging Time	Primary Standards	Secondary Standards
TSP	24 h [1]	240 μg/m³	150 μg/m³
	Geometric Annual Average	80 μg/m³	60 μg/m³
PM10	24 h [1]	150 μg/m³	150 μg/m³
	Arithmetic Annual Average	50 μg/m³	50 μg/m³

[1] Do not exceed once a year.

Table 2. Critical air pollution scenarios, CETESB (2004).

Pollutant	Attention	Alert	Urgency
TSP ($\mu g/m^3$)–24 h	≥375	≥625	≥875
PM10 ($\mu g/m^3$)–24 h	≥250	≥420	≥500

Table 3 presents different country air quality standards for PM2.5.

Table 3. Air quality standards for PM2.5.

Countries/Regions	Averaging Time	Standard ($\mu g/m^3$)	References
WHO	24 h	25	[25]
	Annual	10	
Canada	24 h	30	[26]
USA	24 h	35	[27]
	Annual	15	
European Union	Annual	25 (in 2010)	Directive 2008/50/CE [28]
	Annual	20 (in 2015)	
Mexico	24 h	65	[26]
	Annual	15	

The World Health Organization [25] recommends the standards for PM2.5 air quality as 25 $\mu g/m^3$ using a 24-h averaging time, based on well-known short- and long-term health effects. Canada currently uses two sets of air quality standards: National Ambient Air Quality Objectives (NAAQOs) and Canada-Wide Standards (CWS). The federal government sets the NAAQOs [29]. These are national goals intended to protect public health, the environment and the aesthetic properties of the environment. They take a long-term risk reduction approach to protecting the environment and public health while recognizing economic and technical limits. The goals, which were developed for SO_2, NO_2, CO, O_3 and TSP, are intended to provide background information, a uniform scale for assessing air quality in Canada and guidance to governments. The Canada-Wide Standards (CWS) establish numeric goals for ambient PM2.5 (30 $\mu g/m^3$ using a 24-h averaging time) and O_3 concentrations (65 ppb over an 8-h averaging time) that should be met by 2010 [29].

The Clean Air Act was amended in 1990 and requires that the EPA approves the Standard or National Standard of Air Quality (NAAQS), for compounds considered dangerous to human health and the environment. The Clean Air Act established two types of European Union national air quality standards: the Primary Standard, which enforces standards to define population health, including the health of sensitive populations, such as asthmatics, children and the elderly; the Secondary Standard, which imposes limits to protect public welfare, including protection against decreased visibility, damage to animals, crops, vegetation and buildings [27].

Directive 2008/50/EC (Council of the European Union) [28] became effective in June 2010, and the Member States are responsible for its implementation. This Directive is part of the "Fifth Action Program for the Environment of the EC" of 1992, which recommended the establishment of long-term goals for air quality. It complements European legislation in the field of air quality improvement that involves Directive 80/779/EEC on the air quality standards and reference values for sulfur dioxide and suspended particulates. In addition to the criteria pollutants covered by the NAAQS, the State has promulgated ambient air quality standards for total suspended particulates (TSP), hydrogen sulfide and added a 24-h NO_2 standard [30].

The State of New Mexico air quality regulations are provided in the New Mexico Administrative Code (NMAC) Title 20, Chapter 2. These regulations establish State Ambient Air Quality Standards (NMAAQS) that are equal to or more stringent that the NAAQS. New Mexico also requires that all pollutant concentrations are expressed in parts per million (ppm) and are adjusted for altitude and temperature at the measurement location [30].

The Brazilian regulations do not have standards for PM2.5, while the American regulations do not have standards for suspended particulates. Both regulations do not establish air quality, which takes into account the generation source and the chemical composition of the material, and only establish the weight and the potential toxic risk.

2.3. Equipment Used to Sample Particulate Matter

The sampling of particulate matter can be carried out by different types of equipment. This paper only presents the MiniVol equipment, which was used in the study.

The MiniVol Portable Air Sampler was developed jointly by the U.S. Environmental Protection Agency (U.S. EPA) and the Lane Regional Air Pollution Authority in an effort to address the need for portable air pollution sampling technology. Airmetrics (Springfield, OR, USA) manufactures the MiniVol™ TAS, which samples ambient air at 5 L/min for particulate matter (PM10, PM2.5 and TSP). While not a reference method sampler, the MiniVol™ TAS gives results that closely approximate data from federal reference method samplers. Lightweight and portable, the MiniVol™ TAS is ideal for remote areas or locations where no permanent site has been established [31].

The MiniVol™ comes equipped with low flow and low battery shut-offs and operates from a rechargeable, lead-acid battery. The battery can power the sampler for 24 h of continuous sampling before the battery pack must be exchanged for a freshly charged one. The MiniVol™ also features a seven-day programmable timer, an elapsed time totalizer and rugged PVC construction [31].

In the particulate matter sampling mode, air is drawn through a particle size separator and then through a filter medium. Particle size separation is achieved by impaction. Inlet impactors are available with a 10 micron cut-point (PM10) and a 2.5 micron cut-point (PM2.5). Operating the sampler without an impactor allows for the collection of total suspended particulate matter (TSP).

Critical to the collection of the correct particle size is the correct flow rate through the impactor. To assure a constant flow rate through the size separator at differing air temperatures and atmospheric pressures, the sampler must be adjusted for each sampling project [31].

3. Research Methods

The research strategy adopted in the present work was an experiment, consisting of a planned procedure, starting from a problem and hypothesis, which aims to understand phenomena under controlled conditions, to observe and, also, analyze its results [32,33].

The data were obtained by collecting PM2.5, PM10 and TSP via the MiniVols equipment, selected for this work due to its portability. It was installed during the three different construction site phases, earthworks (Phase 1), superstructure (Phase 2) and finishing (Phase 3), over a 10-day period for each phase in a residential building construction site in Salvador, Bahia, Brazil. In this study, the adjustment of the flow rate was carried out according to the MiniVol Operation Manual [31], and, so a regression curve was plotted to determine the initial flow rate of each MiniVol equipment used in the study.

Furthermore, the sedimentation rate was collected using an adhesive "sticky pad", which consisted of the placing of these adhesives in the neighborhood for 3 days in a horizontal position in order to identify the emission impact of the activities that emit particles in the neighborhood.

In order to monitor the environmental conditions in which this PM was emitted, the Davis Vantage Vue weather station was used. The weather station was installed 5 days before the MiniVols accommodation and was allocated at the exit of the construction site, taking into consideration the regions' prevailing wind, and from the data analysis, the correct sampling position was set in order to attain accurate data for each measurement.

Directional criteria have been suggested in order to monitor the particulate material, useful for reviewing critical emissions in neighboring buildings after the implementation of a construction site. This technique involves the use of sampler pairs placed in opposite positions along the prevailing wind direction. One set is placed at the construction site entrance (measuring the contribution of PM emissions entering the construction site), and the other is situated along the same direction, at the end of the construction site (measuring the emission contribution from the construction site) (Figure 1). The measurements were performed at the same time in two points, following two schedules: during the operation of the construction site (7 am–3 pm), measurements of about 8 h, and during the night shift (5 pm–3 pm), measurements of 22 h.

Initially, a diagnosis through on-site observation was carried out seeking to identify emissions risks, dust-generating activities during the different construction phases, including the observation of materials that were used in the production process, and the technologies used to execute the construction process.

Moreover, the measurement process followed a protocol of data collection, developed in order to standardize procedures and occasional variability, measure the pollution concentrations of particulate matter and evaluate the effects caused by the emission of these particles on the neighborhood. These findings were compared with the standards established by national and international rules.

The chemical and physical analysis of particles were carried out in a laboratory. The physical analysis was performed through the Mettler Toledo analytical balance from to LAPAT (Laboratório de Análise dos Processos Atmosféricos)/IAG (Instituto de Astronomia, Geofísica e Ciências Atmosféricas)/USP (University of São Paulo), used in order to obtain the sampled filter mass and the gravimetry. The concentration of particulate matter measured by a sampler is given by Equation (1). The concentration was calculated for PM2.5 8 h, PM10 8 h, TSP 8 h and PM10 22 h.

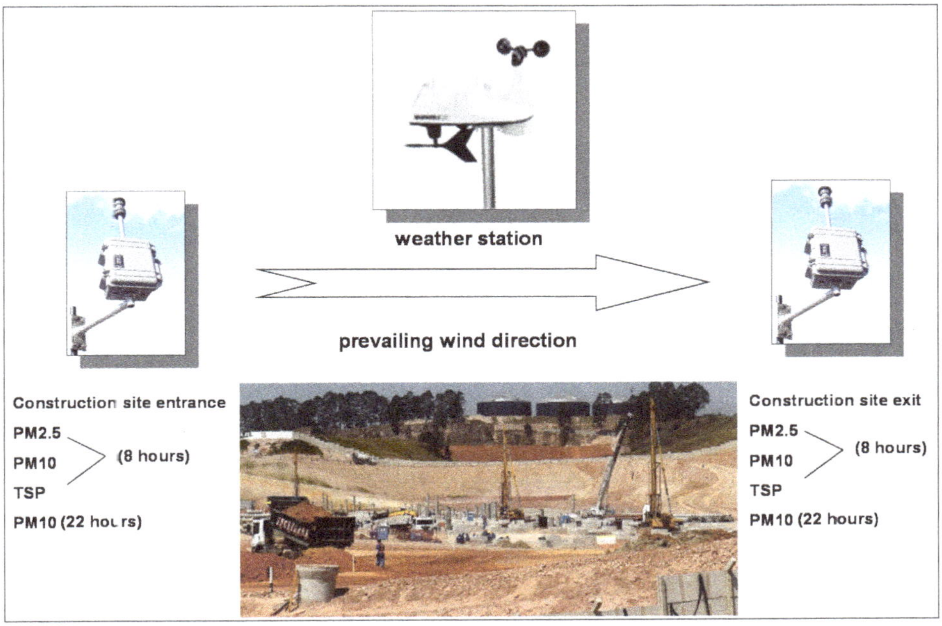

Figure 1. Schematic positioning of equipment at a construction site. Adapted from [34].

Equation (1). Particulate concentration formula:

$$C = m/QxT \tag{1}$$

where:

C = particulate concentration (mass/volume);

m = net mass collected on the filter or substrate (mass);

Q = volumetric flow rate of the sampler (volume/time);

T = duration of sampling (time).

The chemical analysis was accomplished by the X-ray florescence equipment, EDX 700 HS "Energy Dispersive X-Ray Spectrometer" model, from LAPAT/IAG/USP Shimadzu, which was used to quantify the periodic table elements with an atomic number from 11 to 82 (Na to Pb). Thereafter, the results derived from this analysis were converted into concentrations of each identified element by the WinQXAS (Windows Quantitative X-ray Analysis System) program, which aims to improve the data analysis of the X-ray florescence (XRF) through spectral line adjustments. This software allows the manipulation, visualization and simulation of spectral data obtained in the EDX and aims at determining the peak of the interested lines used in the concentration analysis. The X-ray florescence was performed for PM2.5 8 h, PM10 8 h and PM10 22 h, since the material of the filter was a polycarbonate membrane by Whatman. The type of TSP material filter (Teflon by Millipore) does not allow performing the X-ray fluorescence.

The sedimentation rate analysis through the reading of stick pads aims at identifying the percentage of the effective area coverage (EAC) per day. The EAC is calculated by subtracting the instrument reading from the reference area reading and then dividing by the number of days of exposure to give %EAC/day. Table 4 presents the typical levels of response for the reading of stick pads [35].

Table 4. Typical levels of response Hanby (2014). EAC, effective area coverage.

%EAC/day	Response
0.2	Noticeable
0.5	Possible complaints
0.7	Objectionable
2.0	Probable complaints
5.0	Serious complaints

Characterization of the Construction Site

The construction site studied is located in Salvador, Bahia, Brazil, (latitude 12°57′46‴ south, longitude 38°24′32″ west) at an altitude of 34 m. It has an area of 32,780 m^2, with the construction of 8 residential towers, each with 16 floors, totaling 464 housing units.

The construction site is located in a residential urban area (up to 2 floors) with the presence of flora and fauna, including a lake. Within an area of 100 m, there is no presence of primary pollution sources, such as the presence of other construction sites, industries, traffic routes and airports.

Although this study proposed to measure PM during the different construction phases, because the selected construction site had all construction phases in development simultaneously, this was not possible. During the measurement process, overlapping of the phases and consequent mixing of PM from them was observed, thus limiting the ability to analyze the influence of each construction phase and their emitting activities separately. Therefore, an adjustment of the methodology was required. The entry point was considered as a Fixed Point, since it was not shifted along the three different measurement phases, and the exit points were called Point 1, Point 2 and Point 3.

As already mentioned, a weather station was installed 5 days before the installation of the equipment. It is important to note that, depending on the season, time and hour, Salvador's city wind ranges 360°. From the weather station data, the prevailing wind direction found was the southeast wind, especially from 7 am to 3 pm.

The equipment was installed in a fixed point at the main entrance of the construction site and in three distinct points (Point 1, Point 2 and Point 3), located at the construction site exit in different measurement phases. The MiniVols were allocated next to the activities that exhibited a predominance of the main phases of the construction site: Phase 1, earthworks (Point 1); Phase 2, superstructure (Point 2); and Phase 3, finishing (Point 3) (Figure 2).

Figure 2. Position of (**a**) fixed point and Points (**b**) 1; (**c**) 2 and (**d**) 3.

Data collection was carried out as follows:

- *Phase 1*—Measurement of Fixed Point and Point 1 held on the dates: 11/06/2013, 11/07/2013, 11/08/2013, 11/11/2013, 11/12/2013, 11/13/2013, 11/14/2013, 11/15/2013, and 11/19/2013. The sticky pads were installed in the neighborhood for 72 h at 5, 10, 20, 50 and 100 m.
- *Phase 2*—Measurement of Fixed Point and Point 2 held on the dates: 01/07/2013, 01/08/2014, 01/09/2014, 01/10/2014, 01/13/2014, 01/14/2014, 01/15/2014, 01/16/2014, and 01/17/2014. The sticky pads were installed in the neighborhood for 72 h at 5, 10, 20, 50 and 100 m.
- *Phase 3*—Measurement of Fixed Point and Point 3 held on the dates: 01/20/2014, 01/21/2014, 01/22/2014, 01/23/2014, 01/24/2014, 01/27/2014, 01/28/2014, 01/29/2014, and 01/30/2013. The sticky pads were installed in the neighborhood for 72 h at 5, 10, 20, 50 and 100 m.

4. Results and Discussion

This section presents the findings obtained for the three measurement periods (Phase 1, Phase 2 and Phase 3), divided into physical characterization, chemical characterization and sedimentation rate.

4.1. Physical Characterization of Points 1, 2, 3 and the Fixed Point

The samples collected during the construction site activities accumulated an extensive database, acquired by the MiniVols equipment (220 samples). However, the 22-h period of measurement at Points 1, 2 and 3 is not presented in this work due to sampling errors, since the gravimetry values found in their filters showed much lower values compared to the values for 8 h. The inconsistences in these findings could be caused by field measurement errors, laboratory weight procedure errors or equipment procedure errors. Due to the complexity of the exploratory study and because the physical and chemical analysis were performed after all three phase measurements, the authors could not precisely identify the reasons for the sampling errors. Therefore, only the fixed point 22-h period data was used for different measurement phases.

Table 5 presents the descriptive statistics of PM concentrations in μg/m^3 at Points 1, 2 and 3 at Phases 1, 2 and 3, respectively. Table 6 presents the descriptive statistics of PM concentrations in μg/m^3 at the fixed point at Phases 1, 2 and 3. Table 7 presents the weather conditions data per measurement day at Phases 1, 2 and 3. Table 8 presents the summary of the construction activities and dust control methods at Point 1, Point 2, Point 3 and the fixed point. Figures 3–5 present the daily concentration results of PM concentrations in μg/m^3 at Points 1, 2 and 3, respectively.

Table 5. Descriptive statistics of PM concentrations in μg/m^3 at Points 1, 2 and 3.

Descriptive Statistics	Point 1, Phase 1			Point 2, Phase 2			Point 3, Phase 3		
	TSP	PM10	PM2.5	TSP	PM10	PM2.5	TSP	PM10	PM2.5
Maximum	578.86	298.84	101.91	664.19	270.70	105.84	331.13	156.91	50.96
Average	462.25	198.09	75.89	483.12	213.94	77.85	212.31	90.00	46.84
Median	454.32	165.48	75.15	485.60	215.37	71.69	206.35	88.50	46.24
Minimum	361.34	121.52	55.77	307.65	138.28	53.24	95.63	39.73	42.89
Standard Deviation	82.81	71.69	16.74	119.17	50.88	17.19	82.48	39.43	3.21
Coefficient of Variation	18%	36%	22%	25%	24%	22%	39%	44%	7%

Table 6. Descriptive statistics of PM concentrations in μg/m^3 at the fixed point.

Descriptive Statistics	Fix Point, Phase 1				Fix Point, Phase 2				Fix Point, Phase 3			
	TSP	PM10 8 h	PM10 22 h	PM2.5	TSP	PM10 8 h	PM10 22 h	PM2.5	TSP	PM10 8 h	PM10 22 h	PM2.5
Maximum	448.28	131.11	78.32	48.59	327.88	108.01	63.22	58.15	608.97	117.46	56.86	67.87
Average	339.01	90.71	53.26	43.87	287.47	89.31	51.77	43.93	325.09	78.30	46.96	50.24
Median	333.54	87.82	41.16	44.55	297.46	88.97	43.26	42.64	274.71	74.12	33.22	48.13
Minimum	183.30	64.18	48.52	38.56	222.33	60.74	53.24	30.08	125.37	45.43	48.33	38.66
Standard Deviation	90.65	23.96	13.38	3.40	37.65	16.00	7.08	8.48	145.30	23.12	8.88	9.74
Coefficient of Variation	27%	26%	25%	8%	13%	18%	13%	19%	44%	29%	19%	19%

By analyzing Table 5 and Figures 3–5, it is possible to notice that the TSP showed the highest concentrations in the three different phases with average concentrations of 462.25 μg/m^3, 483.12 μg/m^3 and 212.31 μg/m^3, at Point 1, Point 2 and Point 3, respectively.

Besides, the data showed that Point 2 was the most environmentally impacted by concentrations of particulate matter for the three particle sizes (TSP, PM10 and PM2.5) during the studied period when compared with Point 1 and Point 3. Point 2 emitted TSP, PM10 and PM2.5 with average concentrations of 483.12 μg/m^3, 213.94 μg/m^3 and 77/85 μg/m^3, respectively. At Point 1, the average concentrations were 462.25 μg/m^3, 198.09 μg/m^3 and 75.89 μg/m^3, and at Point 3, the average concentrations were 212.31 μg/m^3, 90.00 μg/m^3 and 46.84 μg/m^3 (Table 5). It is important to note that Point 1 and Point 2 emitted a similar PM concentration average of the three fractions.

Table 7. Construction activities and dust control methods at Phases 1, 2 and 3.

Activities and Methods	Phase 1	Phase 2	Phase 3
Construction activities at different points	**Point 1 Main activities**: manual excavation, meso structure, razing of auger piles foundations, vehicular traffic on the soil, land transportation, truck traffic at the construction site. **Other activities around**: superstructure, lift masonry, execution of self-leveling, floor, masonry cement, mortar coating execution, treatment for concrete and floor regularization with cement.	**Point 2 Main activities**: execution of reinforced concrete (columns, beams and slabs), lift masonry, mortar execution, and masonry shaft. **Other activities around**: interior and exterior finishes (ceramic coating (wall), mortar coating, marble and granite fillet door), plumbing and electrical installation systems, waterproofing with asphalt blanket and crystallized waterproofing.	**Point 3 Main activities:** external and internal mortar finishing activities, grouting masonry façade, ceramic coating (wall), crystallized waterproofing, countertops marble and granite, lining plasterboard plates, sanding (wall and ceiling). **Other activities around**: execution of reinforced concrete (columns, beams and slabs), lift masonry, mortar execution and masonry shaft.
Construction activities at **Fix Point**	**Main activities**: transportation of material, moving vehicles and the surrounding areas of construction activities in general. **Other activities around** activities of the neighborhood.		
Dust control methods	Sprinkling water to the routes inside of the jobsite Careful traffic route All trucks entering and leaving site were covered	Sprinkling water to the routes inside of the jobsite Careful traffic route All trucks entering and leaving site were covered	Sprinkling water to the routes inside of the jobsite Careful traffic route All trucks entering and leaving site were covered Netting system around the tower to mitigate dusty façade mortar activities

Table 8. Weather Condition data per day of measurement at Phase 1, Phase 2 and Phase 3.

Phases/ Dates	Humidity (%)	Atmospheric Pressure (mmHg)	Temperature (°C)	Wind Speed (m/s)	Pluviometry	Prevailing Southeast Wind (% per day)
Phase 1						
11/06/2013	69	755.8	24.4	0.9	0.0	11.1
11/07/2013	77	756.1	26.4	1.3	1.4	44.4
11/08/2013	69	758.8	29.2	1.4	0.0	0.0
11/11/2013	66	758.2	30.1	1.6	0.0	17.6
11/12/2013	67	758.2	29.9	1.0	0.0	64.7
11/13/2013	74	758.3	28.9	1.9	0.0	29.4
11/14/2013	72	759.4	28.7	2.3	1.0	0.0
11/15/2013	74	760.0	27.8	1.6	0.2	35.3
11/19/2013	76	759.6	27.8	4.7	0.2	0.0

Table 8. *Cont.*

Phases/ Dates	Humidity (%)	Atmospheric Pressure (mmHg)	Temperature (°C)	Wind Speed (m/s)	Pluviometry	Prevailing Southeast Wind (% per day)
Phase 2						
01/07/2014	77	756.3	26.4	1.3	1.4	44.4
01/08/2014	80	755.5	25.9	1.8	0.0	22.2
01/09/2014	82	755.1	25.0	0.9	0.2	11.1
01/10/2014	89	756.3	25.2	0.9	3.2	0.0
01/13/2014	78	758.5	27.0	1.4	0.0	0.0
01/14/2014	78	758.2	27.4	1.4	0.0	22.2
01/15/2014	83	756.7	26.5	1.5	1.8	0.0
01/16/2014	73	756.8	27.0	1.0	0.0	33.3
01/17/2014	86	757.0	25.4	1.1	1.8	0.0
Phase 3						
01/20/2014	77	756.5	27.8	1.2	0.0	0.0
01/21/2014	76	759.4	28.8	0.6	0.0	0.0
01/22/2014	81	759.8	27.7	0.7	6.6	17.6
01/23/2014	71	759.5	29.1	0.7	0.0	5.9
01/24/2014	76	758.4	26.8	1.4	0.0	0.0
01/27/2014	75	760.1	29.7	1.1	0.0	0.0
01/28/2014	76	758.6	27.9	1.1	1.8	0.0
01/29/2014	72	758.5	29.4	2.0	0.0	0.0
01/30/2014	66	758.9	29.6	1.7	0.0	0.0

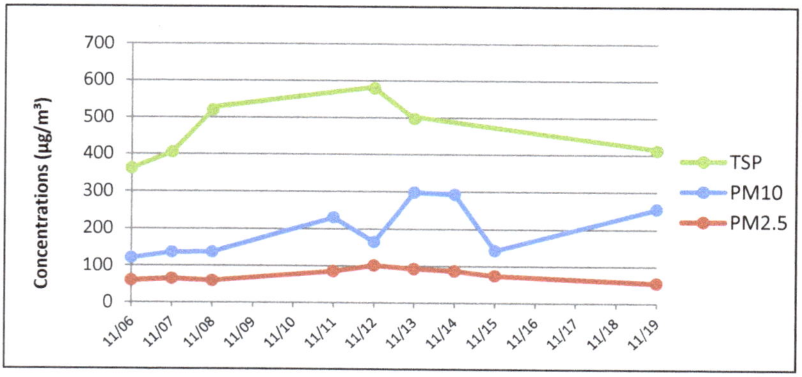

Figure 3. Daily concentration results of TSP, PM10 and PM2.5 in µg/m^3 at Point 1 (7 am–3 pm).

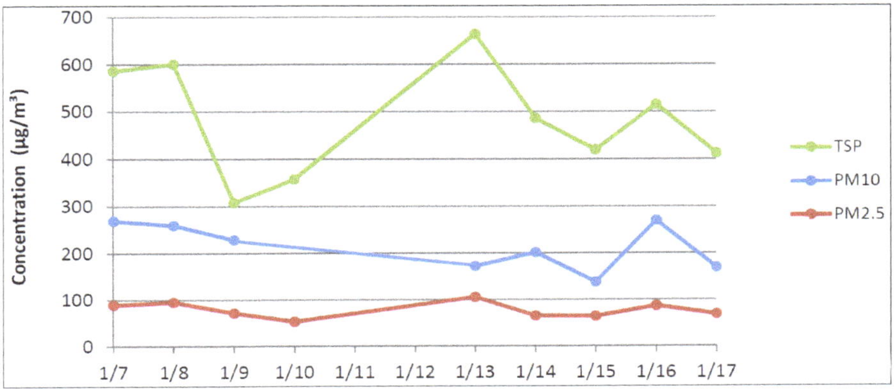

Figure 4. Daily concentration results of TSP, PM10 and PM2.5 in µg/m^3 at Point 2 (7 am–3 pm).

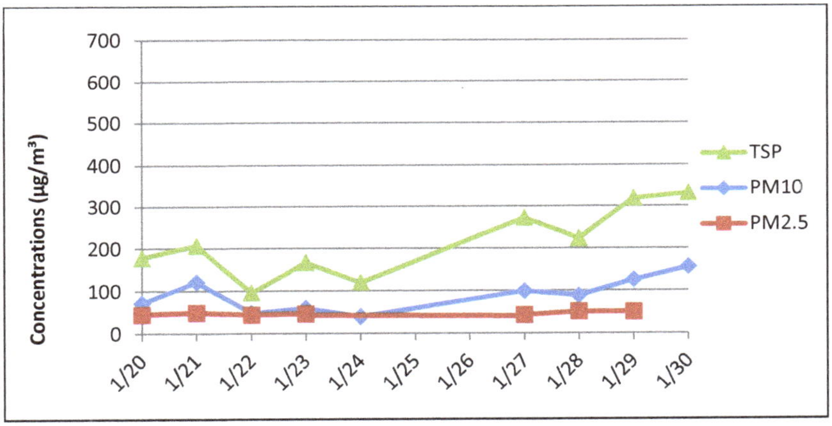

Figure 5. Daily concentration results of TSP, PM10 and PM2.5 in µg/m^3 at Point 3 (7 am–3 pm).

Concerning the fixed point, it is possible to observe that there is less of a difference between average PM concentration (TSP, PM10 8 h, PM10 22 h and PM2.5) data from the different measurement phases (Phase 1, Phase 2 and Phase 3) (Table 6). In addition, there is not a correlation between PM concentrations from the fixed point to the other points at the same measurement phase. This occurred because the fixed point was installed at the entrance of the construction site and was not under the same PM influence as Point 1, Point 2 and Point 3 at Phase 1, Phase 2 and Phase 3, respectively. This can be justified by the large size of the construction site. From the fixed point to Point 1, the distance was 195.8 m; to Point 2, it was 132.34 m, and to Point 3, it was 210.64 m. The construction activities identified at this fixed point were almost the same for the three measurement periods (Table 7) and were related to transportation of material, moving vehicles and the surrounding areas of construction activities in general. Moreover, the particles collected at the fixed point may have been derived from the activities occurring in the neighborhood.

By analyzing Table 8, it is possible to understand the weather conditions during the three measurement periods. The humidity was higher than 60%, with low variation along each phase. The range of atmospheric pressure (755–760 mmHg) and of the average temperature (Phase 1: 24–30 °C; Phase 2: 25–27 °C; and Phase 3: 27–29 °C) had low variation during the three measurement phases. Concerning pluviometry, on 60% of the measurement days, the value found was 0 or 0.2 mm; on one day, it was 3.2 mm in Phase 2, and on another day, it was 6.6 mm in Phase 3. It is possible to assume that for this study, the low variation of humidity, temperature and atmospheric pressure had a low influence on the PM concentration.

Most samples were taken during calm (0–0.2 m/s), light air (0.3–1.5 m/s) and light breeze (1.6–3.3 m/s) wind speed conditions apart from one isolated situation (11/19/2013) taken under gentle breeze (3.4–5.4 m/s) wind conditions, which was according to the Beaufort scale [36]. Through the data, it was observed that most of the maximum PM concentration was associated with higher wind speeds. Therefore, this suggested that that wind speed had a major influence on PM concentration in this study compared to the other weather condition variables.

It is important to note that using a statistical regression curve, no significant correlation was found between any weather condition variable and PM concentration for this study.

Aiming at better understanding regarding the highest and lowest PM concentration values obtained, the meteorological data during the study were analyzed, focusing on these days (Table 8).

The highest concentration of TSP (664.19 μg/m^3) was recorded at Point 2 on 01/13/2014, and on this day, there was the absence of the southeast wind; the average speed wind was 1.4 m/s; the humidity was 78%, and there was no rain. The lowest concentration of TSP at Point 2 (307.65 μg/m^3) was obtained on 01/09/2014 with a prevailing southeast wind of 11.1% compared to the winds of the day, with an average speed of 1.1 m/s; the humidity was 82%, and 0.2 mm of rain accumulated.

It was observed that on the day of the highest concentration, the environmental condition could not be guaranteed to be related to the concentration level found, because the meteorological data obtained on this day tend to favor a low concentration level, except the average of the wind speed. However, there is an assumption that there was a contribution of prevailing winds from other winds directing the PM to the device, adding to higher PM emissions due to construction activities on this day. Regarding the day presenting the lowest PM concentration, the obtained value was expected, due to the southeasterly wind, high humidity and presence of rain during this day.

The highest concentration of PM10 (270.70 μg/m^3) was recorded at Point 2 on 01/16/2014 with the prevailing southeast wind of 33.3% compared to the winds of the day with an average speed of 1.0 m/s; the humidity was 73% (the lowest humidity recorded in the period), and there was no rain. The lowest concentration of PM10 at this Point (138.28 μg/m^3) was obtained on 01/15/2014, and on this day, the southeast wind was absent; the humidity was 83%, and 1.8 mm of rain fell.

Regarding environmental conditions, it was observed that on the day of the highest concentration, there was an average prevalence of southeast wind that tended to direct the PM, together with high humidity, even with the absence of rain. On the day of the lowest concentration, the weather conditions showed no southeast wind and high humidity, and there was the presence of rain. Therefore, the meteorological data are consistent with both findings.

The highest concentration of PM2.5 (105.84 μg/m^3) was recorded at Point 2 on 01/13/2014 (the same day of the highest TSP concentration), in which there was the absence of the southeast wind; the average speed wind was 1.4 m/s; the humidity was 78%, and there was no rain. The lowest concentration of PM2.5 at this Point (53.24 μg/m^3) was obtained on 01/10/2014, and there was no southeast wind; the average speed wind was 0.9 m/s, the humidity was 89% (highest humidity recorded in the period), and the rain accumulation was 2 mm.

As already mentioned, it was observed that on the day (01/13/2014) of the highest concentration, the environmental condition could not be guaranteed to be related to the concentration level found, because the meteorological data obtained on this day tend to favor a low concentration level, except the wind speed. However, there is an assumption that there was a contribution of other prevailing wind directions that directed the PM to the device, adding to higher PM emissions due to construction activities on this day. Regarding the day of the lowest concentration, the obtained value was expected, due to the presence of a southeast wind, high humidity and the occurrence of rain on this day.

According to the data collected, it is possible to affirm that the construction site activities influenced the environment through a higher emission of TSP during the studied period. TSP are particles that have lower permanence in the atmosphere, therefore depositing near the emission sources; that is, near the immediate neighborhood.

The analysis of PM10 concentrations (22 h) at the fixed point at Phase 1, 2 and 3 (Table 6) indicates that the values obtained were always below the concentrations of the primary and secondary standards of the Brazilian CONAMA Resolution Number 3 [24] (150 μg/m^3–24 h). The highest value was 78.32 μg/m^3, which was measured on the following days, 18 and 19 November 2013, of Phase 1. However, data from this research sometimes exceeded the most stringent data of WHO [25], in relation to the Interim Target (IT-3) for 24 h, 75 μg/m^3. Brazilian standards are outdated if compared to the standards established by WHO and international standards.

4.2. Chemical Characterization of Points 1, 2, 3 and the Fixed Point

From the chemical analysis, a similar behavior was observed between the phases that can be better visualized in Figures 6–10. It was not possible to identify the specific activity that emitted these PM, due to the fact that the activities on the construction work site (earthmoving, reinforcing concrete, lifting of masonry, plastering, among others) were being conducted at the same time as the measurement period. In addition, these activities used materials with similar chemical profiles.

At Points 1, 2 and 3, the presence of the same elements was observed, independent of the measurement phase, and the more expressive presence of the elements Ca, Si, Cl, Al and Fe in proportions that follow similar performances (Figures 6 and 7). Sometimes, there is the presence of a few peaks in relation to other phases, such as Ca, that for PM2.5 presented a concentration of 22% at Point 1, 37% at Point 2 and 26% at Point 3. For PM10, a Ca concentration of 20% was observed at Point 1, 31% at Point 2 and 26% at Point 3.

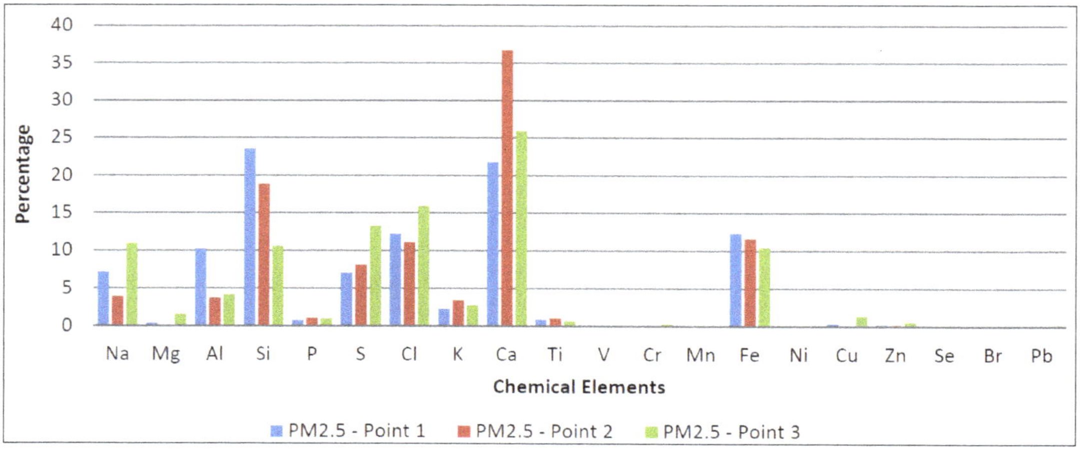

Figure 6. PM2.5, percentage of chemical elements (Points 1, 2 and 3).

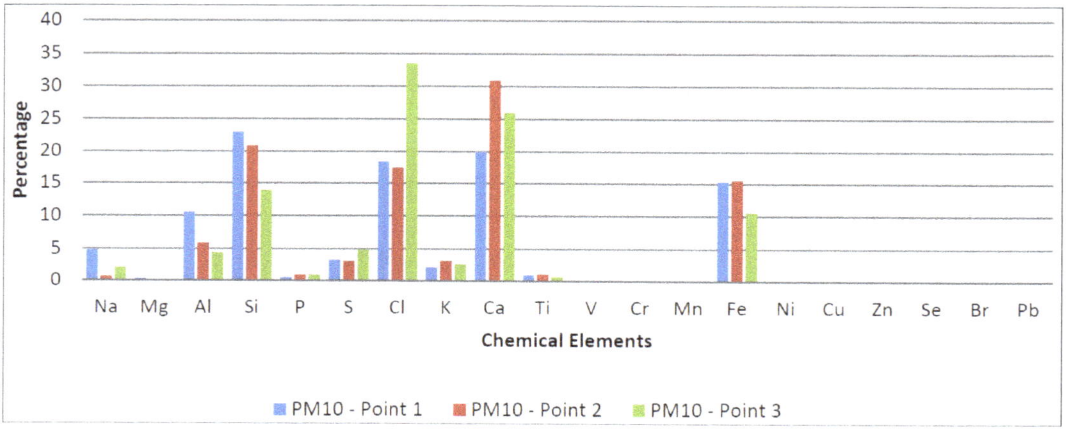

Figure 7. PM10, percentage of chemical elements (Points 1, 2 and 3).

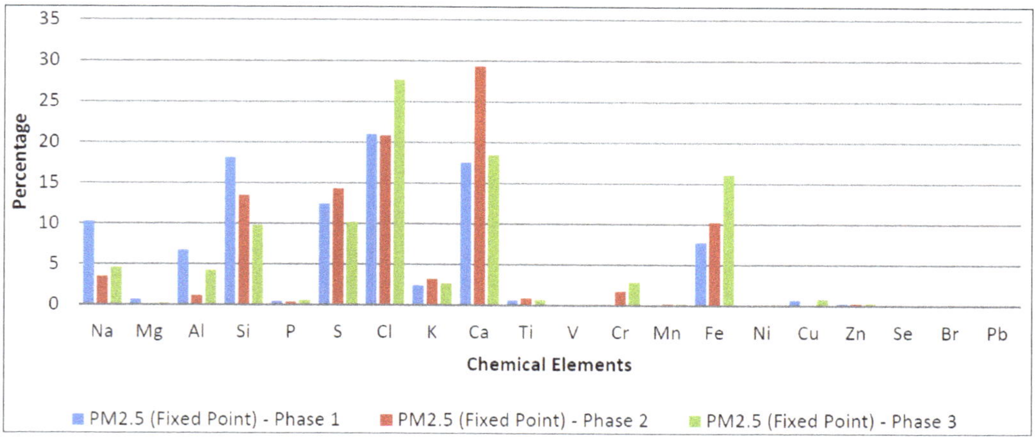

Figure 8. PM2.5, percentage of chemical elements (fixed point).

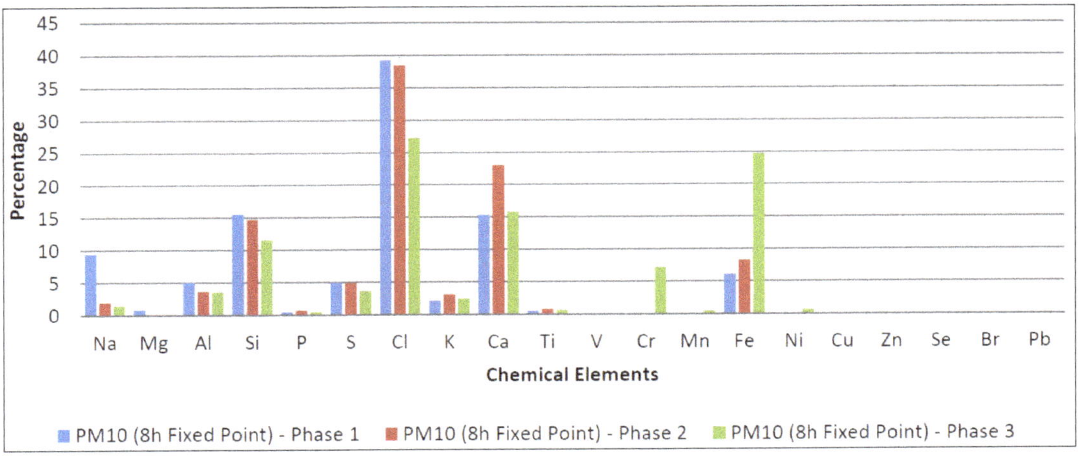

Figure 9. PM10 (8 h), percentage of chemical elements (fixed point).

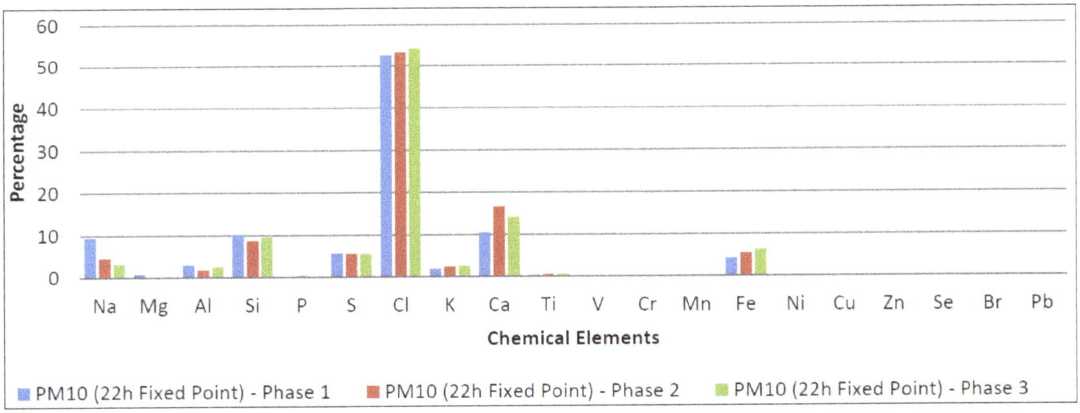

Figure 10. PM10 (22 h), percentage of chemical elements (fixed point).

The chemical data of PM2.5 and PM10 of the fixed point was not different between Points 1, 2 and 3 in which the presence of the same elements was observed, independent of the measurement phase, with a more expressive presence of the elements Ca, Si, Cl, Al and Fe (Figures 8 and 9).

Concerning the chemical data of the PM10 (8 h) and the PM10 (22 h) of the fixed point, again, there were no significant differences, taking into account the presence of the same elements independent of the measurement phase and the more expressive presence of the elements Cl, Ca, Si, Al, Fe and Na. Comparing the graphics of Figures 9 and 10, an enrichment of the elements Cl and Na during the night period (PM10–22 h measurement) was observed, as had been predicted, which can be directly associated with the marine aerosol.

4.3. Sedimentation Rate

The sticky pad results showed that the neighborhood is susceptible to emissions from the construction activities or neighborhood activities at its worst level, since the results presented "serious complaints" in all collections, except at Phase 3 for the measurements of 50 m, which presented probable complaints (see Table 9). However, there is no guarantee that all PM adhered on the stickers had come from the construction activity, since they could receive some contributions from surrounding neighboring areas.

Table 9. Sticky pad reading response.

Meters/Phase	Reference Area Reading	Instrument Reading	EAC	%EAC/day	Response
Phase 1					
5	87	67	20	6.67	Serious complaints
10	87	69	18	6.00	Serious complaints
20	90	74	16	5.33	Serious complaints
50	91	75	16	5.33	Serious complaints
100	93	79	14	4.67	Serious complaints
Phase 2					
5	92	75	17	5.68	Serious complaints
10	96	76	20	6.67	Serious complaints
20	92	76	16	5.33	Serious complaints
50	87	74	13	4.33	Serious complaints
100	93	80	13	4.33	Serious complaints
Phase 3					
5	84	72	12	4.00	Serious complaints
10	89	59	30	10.00	Serious complaints
20	90	80	10	3.,33	Serious complaints
50	91	85	6	2.00	Probable complaints
100	92	80	12	4.00	Serious complaints

4.4. Discussion

Based on the analysis of the collected data in three phases using the proposed methodology, the need for adjustments in some of the methodological procedures adopted was observed in order to obtain satisfactory data according to the reality of the construction site.

In order to evaluate and validate the findings collected during the study, some meetings were carried out with two specialists (Rosana Astolfo, laboratory specialist and MSc in Geosciences of the Astronomy, Geophysics and Atmospheric Sciences Institute, University of São Paulo and Jesuino Romano, manager of the Division of Technology Assessment of Air Quality, Company of Technology Environmental Sanitation) in air quality from the environmental agencies and universities, based on their experience in the assessment of the air quality of vehicle emissions, industry emissions, re-suspension and public health.

The advantages and disadvantages of the typology of the construction site selected for the study was an important topic discussed. As already mentioned, the project studied is characterized by its large and complex construction site, in which several construction activities occurred at the same time and over the entire measurement period of the study. It is important to highlight that this construction site was chosen initially because it was possible to validate the methodology in the three different construction phases over a short time period. However, because the construction site has overlapping phases, it was not possible to identify the contribution of each specific construction phase. The suggestion is the installation of MiniVols on smaller construction jobsite areas in order to better control the measurement process and identify if there are specific contributions of PM due to different construction phases.

Regarding the chemical analysis, the X-ray fluorescence (XRF) analysis technique was used in this study, which provides only chemical elements. The results show that the chemical profile of the PM collected was similar among them, since it showed almost the same chemical elements. Therefore, it was not possible to specify the percentage contribution of each construction phase in the PM in this study. However, other techniques can be used in such studies, such as X-ray diffraction, which provides the chemical compounds. In this research, X-ray diffraction analysis was not used, because the size of the filter does not allow the collection of sufficient material for such analysis.

The measurement of sedimentation rates performed with the sticky pads provided insufficient data for evaluating the impact on the neighborhood, because the PM captured on the stickers came from the construction activities and the surrounding neighboring areas, instead of only the construction site. However, the sticky pad can be tested within the construction jobsite in order to collect the PM emissions from the construction activities, as an alternative and fast method to understand the level of response in terms of PM emission.

5. Conclusions

This work contributes with a methodology to evaluate the emission of particulate matter on construction jobsites from an environmental perspective, taking into consideration different construction phases, despite the limitation pointed out within the paper. This methodology allows for quantifying and characterizing the PM and also enables further intervention through mitigation measures in order to reduce the pollution.

However, corroborating Castelanni [37], particulate emission monitoring is a challenging technical field, not only because of the specific accuracy and performance of particulate monitors, but also due to the harsh environment in which they must continuously operate. Specifically, in the construction industry, the challenge is increased by the complexity and unique characteristics of each construction product and production process. The amount of variables include different construction activities, as well as on-road vehicles associated with the construction site and on machinery, which occurs simultaneously during the whole day without a visible pattern, influencing the emission of PM from different sources. The emission of the PM will also depend on the technology and management control methods used by each project. Finally, the weather condition variables influence the emission concentration of PM.

The paper also shows that the literature about particulate matter from an environmental perspective on construction jobsites is still scarce, so there is much knowledge yet to be explored on this subject. Furthermore, there is a need for more quantitative data concerning the PM emissions in the construction industry, similar to the effort to quantify and establish parameters for construction and demolition wastes over the last decade. There is a need for better understanding about the type, size and concentration of these particles in different construction phases, as well as the impact of this on the neighborhood and construction workers.

The findings show that the activities on construction sites emit different types of particles with environmental impacts through a higher emission of TSP. The concentration values obtained for PM10 22 h in this study did not exceed the limits set by the Brazilian standards established by CONAMA Resolution Number 03; however, some of the level of concentration measured exceeded the standards set by the World Health Organization that represent an increase in short-term mortality.

What is more, this work shows the importance of monitoring the weather condition parameters, such as temperature, humidity and wind direction, in order to gain a better understanding of the behavior of the air pollution. The air quality depends on the quantity of the particulate matters emissions, but also the way in which the atmosphere reacts to their concentration or dispersion. However, due to the exploratory nature of this study, focusing on finding the most suitable methodology for determining PM concentrations around construction sites, and the limited amount of information on the measured concentration, it was not possible to correlate weather condition variables to PM concentration.

In addition, the findings show that Point 2, which was mostly influenced by the superstructure construction activities, was the most environmentally impacted by the PM concentration. However, due to the size and the complexity of the construction site studied, mainly the overlapping of the construction activities during the entire period of the study, there is a need for better investigation concerning the influence of different construction phases on PM emission.

The methodology and results of this research can provide starting points for further studies aimed at measuring particulate matter emissions on construction sites. Some recommendations for future research are:

(1) to develop studies on construction sites to support specific parameters of air quality for the building,
(2) to develop specific studies for each phase of construction, such as earthwork, superstructure and finishing,
(3) to develop specific studies to correlate weather condition variables with PM concentration and
(4) to develop and implement studies to evaluate technological and management solutions for the reduction of PM emissions from construction sites.

Acknowledgments

We are grateful to the Cantechis Project (Technologies for Sustainable Construction on Social Housing Jobsites), Ministry of Science, Technology and Innovation, trough FINEP (Financiadora de Estudos e Projetos) and CNPq (Conselho Nacional de Desenvolvimento Científico e Tecnológico) and the Ministry of Education, CAPES (Coordenação de Aperfeiçoamento de Pessoal de Nível Superior), for the financial support and to LAPAT (Laboratório de Análise dos Processos Atmosféricos)/IAG (Instituto de Astronomia, Geofísica e Ciências Atmosféricas)/USP (University of São Paulo) and CETESB (Company of Technology Environmental Sanitation) for the technical support.

Author Contributions

Dayana B. Costa, Ingrid P. S. Araújo and Rita J. B. de Moraes conceived and designed the study with the support of the Cantechis Project researchers. Ingrid P. S. Araújo performed the data collection and analysis with the support of Rita J. B. de Moraes. Ingrid P. S. Araújo wrote the Portuguese version of the manuscript with the support of Rita J. B. de Moraes. Dayana B. Costa provided academic advice throughout the process, wrote the English version and revised the manuscript.

Conflicts of Interest

The authors declare no conflict of interest.

References

1. Levin, H. Systematic Evaluation and Assessment of Building Environmental Performance (SEABEP). In Proceedings of the Second International Conference of Buildings and Environment Performance, Paris, France, 28–29 November 1997; pp. 9–12.
2. Cho, Y.K.; Alaskar, S.; Bode, T.A. BIM—Integrated sustainable material and renewable energy simulation. In *Construction Research Congress 2010. Innovation for Reshaping Construction Practice*; Ruwanpura, J., Mohamed, Y., Lee, S., Eds.; American Society of Civil Engineers: Banff Alberta, AL, Canada, 2010; pp. 288–297.
3. Tah, J.H.M.; Abanda, H.F. Sustainable building technology knowledge representation: Using Semantic Web techniques. *Adv. Eng. Inf.* **2011**, *25*, 547–558.
4. Cohen, A.J.; Anderson, H.R.; Ostro, B.; Pandey, K.D.; Krzyzanowski, M.; Künzli, N.; Gutschmidt, K.; Pope, A.; Romieu, I.; Samet, J.M.; *et al.* The global burden of disease due to outdoor air pollution. *J. Toxicol. Environ. Health A* **2005**, *68*, 1301–1307.
5. Councils London. *Guidance Note Construction Sites–MODULE 3*; Local Government: Air Quality Toolkit, London, UK, 2010.
6. Resende, F. Atmospheric Pollution Emission of Particulate Matter: Evaluation and Control at Construction Sites of Buildings. Master's Dissertation, University of São Paulo, São Paulo, Brazil, 2007; p. 232.
7. Councils London. *The Control of Dust and Emissions from Construction and Demolition—Best Practice Guidance*; Councils London: London, UK, 2006.
8. Amor, R. The development of sustainable: Construction planning system. *J. Inf. Technol. Constr.* **2012**, *17*, 1874–4753.
9. Greenspan, C.A.; Moure-Eraso, R.; Wegman, D.H.; Oliver, C. Occupational hygiene characterization of a highway construction project: A pilot study. *Appl. Occup. Environ. Hyg.* **1995**, *10*, 50–58.
10. Blute, N.A.; Woskie, S.R.; Greenspan, C.A. Exposure Characterization for Highway Construction Part I: Cut and Cover and Tunnel Finish Stages. *Appl. Occup. Environ. Hyg.* **1999**, *14*, 632–641.

11. Kjell, T.; Bergdahl, I.A.; Nilsson, T.; Järvholm, B. Occupational exposure to particulate air pollution and mortality due to ischaemic heart disease and cerebrovascular disease. *Occup. Environ. Med.* **2007**, *64*, 515–519.
12. Maioli, B.G. Quantification and Characterization of Fine Particulate Matter (PM2.5) in the Metropolitan Region of Vitória, ES. Master's Thesis, Technology Center, Federal University of Espírito Santo, Vitória, Brazil, 2011; p. 118.
13. Daly, A.; Zannetti, P. *An Introduction to Air Pollution—Definitions, Classifications, and History. Chapter 1 of Ambient Air Pollution*; Zannetti, P., al-Ajmi, D., al-Rashied, S., Eds.; The Arab School for Science and Technology (ASST), The Enviro Comp Institute: Freemont, CA, USA, 2007.
14. Schwartz, J.; Zanobetti, A.; Bateson, T.F. Mortality and Morbidity among Eldery Residents of Cities with Daily PM Measurements. In *Revised Analyses of Time-Series Studies of Air Pollution and Health Boston*; Health Effects Institute: Boston, MA, USA, 2003; pp. 25–58.
15. Peters, A.; Pope, C.A., III. Cardiopulmonary mortality and air pollution. *Lancet* **2002**, *360*, 1184–1185.
16. Gualtieri, M.; Mantecca, P.; Corvaja, V.; Longhin, E.; Perrone, M.G.; Bolzacchini, E.; Camatini, M. Winter fine particulate matter from milan induces morphological and functional alterations in human pulmonary epithelial cells (A549). *Toxicol. Lett.* **2009**, *188*, 52–62.
17. USEPA-United States Environmental Protection Agency. *Air Quality Criteria for Particulate Matter*; EPA/600/P-99/002aF; USEPA-United States Environmental Protection Agency: Washington, DC, USA, 2004; p. 900.
18. Environment Agency. *Monitoring of Particulate Matter in Ambient Air around Waste Facilities*; Technical Guidance Document (Monitoring), M17; Environment Agency: Rotherham, UK, 2004. Available online: http: //www.environment-agency.gov.uk (accessed on 8 May 2013).
19. Milford, J.B.; Davidson, C.I. The size of particulate trace elements in the atmosphere. *Rev. J. Air Pollut. Control Assoc.* **1985**, doi:10.1080/00022470.1985.10466027.
20. Kelly, F.J.; Fussell, J.C. Size source and chemical composition as determinants of toxicity attributable to ambient particulate matter. *Atmos. Environ.* **2012**, *60*, 504–526.
21. Seigneur, C. Current status of air quality models for particulate matter. *J. Air Waste Manag. Assoc.* **2001**, *51*, 1508–1521.
22. Sielicki, P.; Janik, H.; Guzman, A.; Namieśnik, J. The progress in electron microscopy studies of particulate matters to be used as a standard monitoring method for air dust pollution. *Crit. Rev. Anal. Chem.* **2011**, doi:10.1080/10408347.2011.607076.
23. CETESB—Company of Technology Environmental Sanitation. Report of São Paulo Air Quality. Available online: http://www.cetesb.sp.gov.br/Air/air_health.asp (accessed on 8 May 2013).
24. CONAMA—National Environment Council. CONAMA Resolution Number 3. Available online: http://www.mma.conama.gov.br/conama (accessed on 8 May 2013).

25. WHO–World Health Organization. Air quality guidelines for particulate matter, ozone, nitrogen dioxide and sulphur dioxide. In *Global Update 2005, Summary of Risk Assessment*; WHO–World Health Organization: Geneva, The Swiss, 2006.
26. Lee, S. Fine particulate matter measurement and international standardization for air quality and emissions from stationary sources. *Fuel V.* **2010**, *89*, 874–882.
27. USEPA—United States Environmental Protection Agency. Particulate Matter (PM) Standards—Review Completed in 2006. Available online: http://www.epa.gov/ttn/naaqs/standards/pm/s_pm_cr.html (accessed on 8 May 2013).
28. European Parliament; Council of the European Union. Directive 2008/50/EC of the European Parliament and the Council of 21 May 2008 on Ambient Air Quality and Cleaner Air for Europe. Available online: http://eur-lex.europa.eu/LexUriServ/LexUriServ.do?uri=OJ:L:2008:152:0001:0044:EN:PDF (accessed on 10 August 2014).
29. Croes, B.; Schneider, J.; Hedley, A. Air quality management approaches and evidence of effectiveness. In *NERAM Guidance Document on Air Quality Risk Management*. Network for Environmental Risk Assessment & Management: Waterloo, ON, Canada, 2007; pp. 99–154.
30. US Department of the Interior, Bureau of Land Management. *BLM Socorro Field Office RMPR/EIS 33 Air Quality Impact Analysis*; US Department of the Interior, Bureau of Land Management: Washington, DC, USA, 2003.
31. Airmetrics. *Minivol Operation Manual*; Airmetrics: Eugene, OR, USA, 2011. Available online: http://www.airmetrics.com (accessed on 20 April 2014).
32. Walliman, N. *Research Methods: The Basics*; Routledge: New York, NY, USA, 2011.
33. Dawson, C. *Introduction to Research Methods: A Practical Guide for Anyone Undertaking a Research Project Paperback*; How To Books Ltd.: Oxford, UK, 2009.
34. Resende, F.; Cardoso, F.F.; Costa, D.B.; Melo, A.; Araújo, I.P.S.; Moraes, R.J.B.; Porfiro, A. Methodology for Measuring the Concentration of Particulate Material in Construction Sites. In Proceedings of the SIBRAGEC 2013—8th Brazilian Symposium of Construction Management and Economics, Innovation and Sustainability. Brazilian Association of the Built Environment Technology (ANTAC), Salvador, Brazil, 18–20 November 2013.
35. Hanby, I. Calibration & Use of the Sticky Pad Reader Guidelines for Use of the Sticky Pad Reader. Available online: http://www.hanby.co.uk/CAL%20&%20USE%20SPR.htm (accessed on 29 September 2014).
36. Huler, S. *Defining the Wind: The Beaufort Scale, and How a 19th-Century Admiral Turned Science into Poetry*; Three Rivers Press, Crown Publishing Group: New York, NY, USA, 2004.
37. Castelanni, B.; Morini, E.; Filipponi, M.; Nicolini, A.; Palombo, M.; Cotana, F.; Rossi, F. Comparative analysis of monitoring devices for particulate content in exhaust gases. *Sustainability* **2014**, *6*, 4287–4307.

Generation of a Tropically Adapted Energy Performance Certificate for Residential Buildings

Karl Wagner

Abstract: Since the 1990s, national green building certification indices have emerged around the globe as promising measurement tools for environmental-friendly housing. Since 2008, tools for countries in the Northern "colder" hemisphere have been adapted to tropical countries. In contrast, the Tropically Adapted Energy Performance Certificate (TEPC), established in 2012, translates the United Nations' triple bottom line principle into green building sustainability (planet), thermal comfort (people) and affordability (profit). The tool has been especially developed and revamped for affordable green building assessment helping to reduce global warming. Hence, by the comparably simple and transparent energy audit it provides, the TEPC examines buildings for their: (1) contribution to reduce CO_2; (2) transmission rate in shielding a building's envelope against the effects of the tropical heat; (3) generation of thermal comfort and (4) referring total cost of ownership to green the building further. All four dimensions are measured in the rainbow colour scale in compliance with national energy regulations. Accordingly, this research examines the tool's implementation in tropical countries. Exemplified tropical case studies in residential areas seek to demonstrate the practicability of the approach and to derive a holistic certification by an internationally accredited certification board.

Reprinted from *Sustainability*. Cite as: Wagner, K. Generation of a Tropically Adapted Energy Performance Certificate for Residential Buildings. *Sustainability* **2014**, *6*, 8415-8431.

1. Introduction: Beyond the Sick Building Syndrome

In recent years, several high-tech developing countries have become sensitive and proactive towards green solutions for the tropical built environment. The "Rainforest Alliance" based in New York is a globally operating body for the tropics. Colombia and Singapore belong to those countries having established green building councils. Over the past decade, several national Green Building certification indices have emerged. The American LEEDS, the German DNGB or the British BREAM are all considered as measurement tools for environmental-friendly housing in the tropical hemisphere. Since 2009, such tools for countries in the Northern "colder" hemisphere have been adapted towards tropical countries (e.g., the Singaporean Green Mark Index). Since Malaysia established its Green Building Index and committed to saving 40% of CO_2 emission by 2020 (it has been spearheaded as a country embarking on viable environmental survival strategies, along with Germany [1]. Countries like these find themselves at a pivotal point to turn green ideas into practice (*cf.* [2]). Due to their much higher electricity costs, the so-called developed countries in the colder hemisphere are far more restricted in their energy consumption. Since 1990, some of them have already devised and monitor a wide variety of energy regulations and accordingly have developed renewable technologies that might be adaptable to green buildings [3].

This contribution focuses on the adaptation and implementation of the Energy Performance Certificate (EPC)—which two decades ago was established by nations in the European Union—for a tropical country in South-East Asia [4]. Occupants are no longer simply victims of the offending "sick building syndrome" [5]. Supporting their need for transparency and simplicity in understanding their building in terms of its vulnerability against the climate, the tool is implemented with an easy to understand scientific tool kit to measure CO_2—including emissions, insulation, thermal comfort and cost considerations. With the assistance of a green building consultant, it sketches a staggered procedure of how to save CO_2 as the first priority. Hence, the proposed and tested toolkit starts with the question of "how green the building is" along the rainbow scale.

We can consider a property developer or a landlord ambitious to sell his or her green and energy efficient tropical building. The person has invested in green measures and enjoys both higher thermal comfort and reduced monthly electricity bills. He or she can claim to be a greener person, contributing in some ways to save the environment. Will the achievement be rewarded when the person attempts to sell the building for more revenue as compared to a landlord who did not go green? At first, let us study to the following simulated conversation between a potential house buyer and the present landlord about a Tropically-Adapted Energy Performance Certificate (TEPC) for residential housing.

Potential Buyer: "Before we negotiate about the price of the terrace house you intend to sell, can I see your TEPC?"

Owner: "My Building was assessed 3 years ago, and you can see it is green in terms of all four criteria. So, the price is about 10% higher than the neighbouring unit, but you will feel comfortable. As you will save 70% of the energy needed to cool it down, your predicted payback timeframe is about 3.5 years."

2. Energy Assessment Tool for Residential Housing

Application of an Energy Demand Certificate (EDC) for new residential buildings and an Energy Performance Certificate (EPC) for existing residential buildings is the current practice within the European Union, with so far at least five countries requiring their implementation for any building. Furthermore, a growing number of countries outside Europe have also adapted these tools. Effective since 2009, e.g., in Italy or Portugal every seller of real estate has to enclose the energy performance certificate as an appendix to the sales agreement (e.g., [6]).The tropical EPC which is outlined within this paper suggests a cautious adaptation to a warm country where cooling is by far the major environmental issue triggering operational costs of the modern built environment. This tool provides no status of platinum, gold or silver. As Figure 1 shows, it rather caters for a detailed scoring-based school mark system from A+ to G aligned with the spectral colours (*cf.* [7]):

Figure 1. Energy Performance Certificate (EPC) for buildings (respective of the thermographic profile) (*cf.* [8]).

Following both the school marks system and the thermographic profile, any building's energy performance can be certified between red and blue. A building which is fully independent or is producing its own renewable energy, or can even can feed electricity into the grid will be granted a mark higher than "A" or green for its energy consumption [9]. The blue A+ building itself-sustainable like all natural life on our blue planet. In their extremes, A+-buildings are power producers that hence might supply the grid under the Feed-In Tariff ("FIT"), e.g., in Malaysia since 2012 [10].

Hence, the practical implication of the TEPC is not a green label solely shown to relatives, neighbours and friends of how "green" and perhaps, also energy-efficient the building is. It is there to assist the occupant and a prospective buyer to gauge to what extent it is to save energy and hereby save CO_2 in the future. In addition, selling will be easier, and the certificate is affordable (following some prevailing European standards the price would be about 100 EUR). The tool is distinct from more sophisticated Green Building Indices such as LEEDS—USA, BREAMS—United Kingdom, Green Star—Australia, DGNB—Germany or GBI—Malaysia [11]. The reason is that the EPC is solely interested in evaluating (a) the strength of insulation and subsequently; (b) the greenness of the electricity consumption in terms of energy efficiency.

Along with two further criteria derived below, the TEPC certificate will show to what extent a building is capable of saving on energy usage and, in the blue case it will even produce renewable energy for the grid and other households. In its radical version as a passive house, it is clearly benchmarked against other reference buildings. The following example shows the overall energy consumption comparing different reference buildings with different standards (retrofitted, newly erected on the scale of the most recent German Energy Regulation 2014 (EnEV).

If a house were to comply with the official regulations for a new building adhering to Energy Regulation (2012), according to Figure 2 the primary energy demand would have an allowable maximum consumption of 100 kWh/m^2 [12]. Furthermore, the "light-yellow" in-part retrofitted building in the yellow zone may consume a maximum 200 kWh/m^{2}p.a. of primary electricity. That would mean that in some cases of existing houses major renovations—like the not retrofitted building with 425 kWh/m²p.a have to be undertaken in order to bring them from a higher kWh/m²p.a. consumption. They enable the building to attain the permitted 200 kWh/m²p.a of an in-part retrofitted building in our example in the subsequent Figure 2). Proven methods include replacing single with double glazing with the effect that the primary energy demand per annum will decrease below the targeted value of 200 kWh kWh/m²p.a—if the green building consultant's recommendation (e.g., consumption is monitored by an energy saving smart system and mechanical night-time cross ventilation) is implemented. Figure 2 exhibits the status of a passive house and energy regulations in Germany benchmarking real buildings scaling their distances.

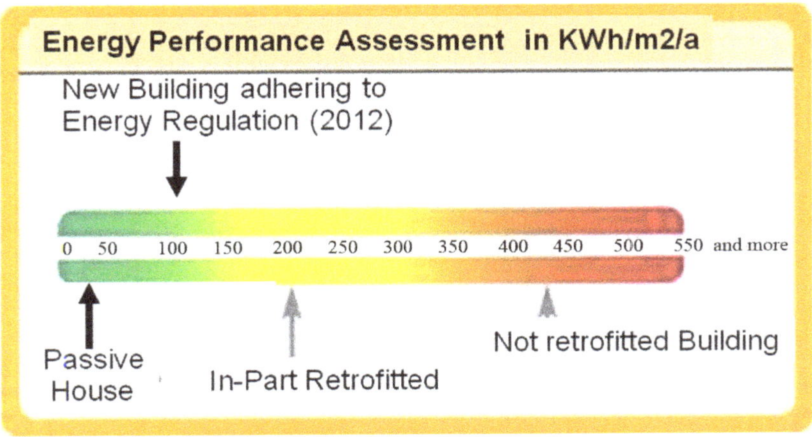

Figure 2. General scale of EPC KWh/m²—summary and examples of yellow and red zone building legal requirements in Germany.

The following depiction shows the fully-fledged energy performance certificate of a retrofitted building near Stuttgart, Germany. The house consumes 12 kWh/m²p.a. of power and has 26 kWh/m²p.a primary energy demand which renders it already quite green. With this performance certificate, when selling the house, the new owner will have an indication of the greenness of the building and this will help to negotiate accordingly towards a reasonably higher market price compared to a common red or yellow building [6]. As pointed out above, the fully-fledged certificate form is an official document obligatory to be presented for a purchase in several countries of the EU.

3. Measuring the Status Quo in a Residential Building

Throughout the following pages, in total, four core dimensions of the Tropically Adapted Energy Performance Certificate (TEPC) will be derived from the Triple Bottom Green Line mentioned above.

CO_2 emission, thermal comfort and cost saving are the three "triple" output-factors posing as a "magic triangle" for green and energy efficient buildings. Conversely, minimum heat transmission (by optimum insulation, up to air-tightness) is considered an enabler or input factor for the three output factors. Basically, the four factors can be individually measured between green and red. The occupant or owner of a building in Figure 3. receives feedback whether these items are green, red or somewhere in between in the rainbow scheme.

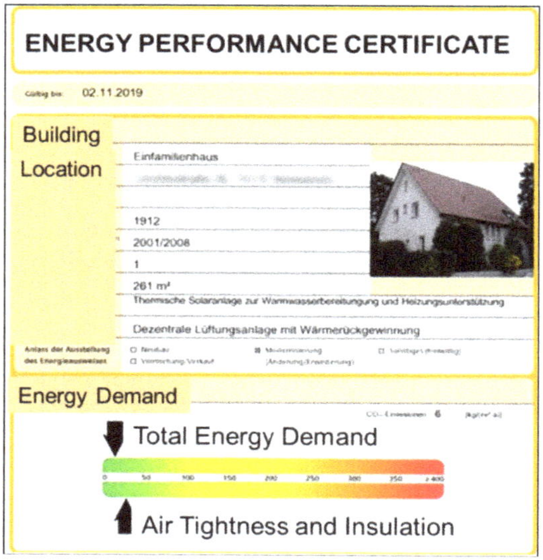

Figure 3. Energy Performance Certificate residential building (Germany) [13].

Instead of measuring two parameters like for the original European tool which has also been commonly transferred to other parts of the world, the TEPC does not only distinguish between CO_2 emission as well as heat transmission and insulation rate. In addition, thermal comfort and cost considerations are introduced as two more parameters of the Tropically Adapted Energy Performance Certificate. Different from countries in the colder hemisphere, where thermal comfort seems to be taken for granted within the certification tools, the achievement of thermal comfort is very questionable especially for non-automated tropical residential buildings [14]. Furthermore, cost consideration is a decisive element to create awareness, reflect and utilise the purchasing power of potential green building buyers. Bringing cost considerations into the fray seems to create new business terrain in a market place where it appears that Returns of Investment (ROIs) comparing green and red had never been calculated in a systematic and meaningful manner.

The following four subunits provide a closer look into the four categories paving the way for further discussion.

3.1. Heat Transmission and Insulation Rate

Within the current measurement, it is a rising imperative and common practice in the green building industry to measure heat rejection indicators such as R-values or summarised *U*-values, A-values, CHI-values and so forth. Within the TEPC, these values are enablers to reduce the basic temperature and try to help to determine the Roof Thermal Transfer Value (RTTV) and the Energy Thermal Transfer Value (ETTV). As a concerted action, however, the approach mobilised here is also quite different and easy to understand for its occupants who might not be engineers or architects. Still, we are utilising materials in terms of heat reduction, but in the end there is only one relevant factor to determine the greenness or redness of a building's insulation. The radiant "surface" temperature of all of the following nine parameters at peak values during a set of five typical sunny/cloudy "overcast" afternoons are considered with the following measurement tool.

According to this template, the radiant temperature is considered as an input factor which enables a building to create a more or less agreeable ambient temperature inside. It is possible to calculate the ambient temperature by using the single parameters above. The precise way most likely lies in mathematical formula yet to be developed. The easier way is to rely on the statistical values of one's own research as we did in case of a room in an urban high rise in comparison to a suburban semi-detached building. The following rule of thumb calculation at this point in time is based upon empirical evidence in the author's own house: The average of the radiant temperature of the room's eight surrounding modules in Figure 4 minus 1 = ambient room temperature at the same point in time. That means if we are able to bring down the radiant average peak day temperature to 29.6 °C, the ambient temperature will be 28.6 °C. This is the allowable maximum in range of the tropical thermal comfort zone devised below [9,15]. If the average radiant temperature of the surroundings is still above 29.6 °C, it would be necessary to cool the respective room module further or to decrease the ambient temperature until it is in range and can be maintained. Therefore, the colour scale along with a more detailed scale from 0–100 scores can serve as an overall TEPC-scoreboard as per Figure 4:

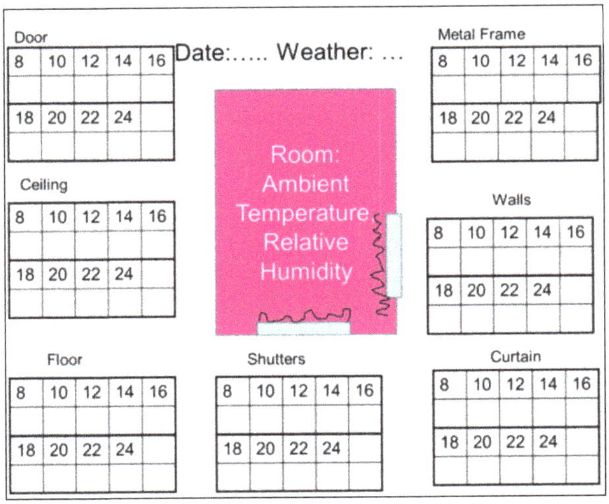

Figure 4. Ambient temperature of a room and radiant temperature of its seven surrounding modules measured nine times daily (in correlation with outside temperature).

In contrast to the A+ value of the blue European energy performance certificate, blue in a tropical sense of surface temperature would mean the delivery of too-cold radiant temperatures. According to a normal distribution elaborated in Section 3.3 under thermal comfort, this temperature has not been requested by the majority of tropical occupants. Therefore, the scores (max. 100, min. 0) are also low, similar as if the surface temperature average exceeds 29.6 °C. Figure 5 presents the application of the TEPC scaling radiant inside temperatures of the building envelope and related parameters like windows and doors.

Celsius	COLOUR	Score for overall Scoreboard (continuous)
< 24	DARK BLUE	0-59
24- 25.9	BLUE	60-79
26-27.7	European GREEN	80-90
27.8-29.6	Tropical GREEN	90-100
29.7-31	ORANGE	79- 50
31-33	RED	50-10
>32	DARK RED	9-0

Figure 5. Input U-/G-value replaced by output radiant temperature of 6–7 parameters [14].

3.2. Carbon Footprint

Within the TEPC, CO_2 in relationship with the operation of electrical gadgets is the only emission threat considered.

In addition, unlike the common Green Building Indices referred to above, the TEPC does not yet include the carbon footprint through the generation and life cycle costs of the building. Especially, the innumerous existing buildings have to be conceived of as a long-term investment that has been constructed before the contractors could implement green and energy efficient features. The possibility that even the initial carbon footprint has been ineradicably made will decline over the course of time. In the case of walls, doors and roofs, natural biodegradable and replantable local and localised materials might be preferable compared with those which have to be produced with a large energy-respective carbon footprint. The usage of recyclable material (like in the case of light-weight concrete or wood wool) is a plus, in case we cannot fully rely on natural materials like paddy husk or palm oil fibre alone. Furthermore, locally manufactured products and services for an energy-efficient green building should be chosen over those from far away or from foreign countries (*cf.* [16]).

If we estimate the life expectancy of a building at 50, 60 or 70 years, the carbon footprint of its generation will be factored in to every year of its operation. In addition, by using as many natural materials as possible, the generation of carbon footprint will decrease.

Waste reduction, recycling green building materials, utilisation of renewable green energy or other indicators are not included either because the TEPC is "just" a CO_2-tool for green and energy efficient buildings. It is a tool which is based on the same admittedly simple but not simplistic international standards of the Northern hemisphere and is restricted to CO_2.It is believed that CO_2 by far poses the biggest threat to survival that can only be tackled by transnational actions. Hence, the extension into life cycle analysis, further gases and other issues mentioned above are undoubtedly welcomed ideas for the future development of the TEPC as a supplementary or stand-alone certification tool.

For its practical operation, two different measurements both accounting for 50% of the overall value of the determined carbon footprint can be distinguished: CO_2/m^3/p.a. and CO_2 per occupant. In the tool shown in Figure 6, we chose "occupant".

CO2/occupant /p.a.	COLOUR	Score for overall Scoreboard
< 100	DARK GREEN	100
100- 129	GREEN	90
130-149	LIGHT GREEN	80
150-169	YELLOW	60
170-190	ORANGE	40
190-209	RED	20
>210	DARK RED	0

Figure 6. Exemplified green–red TEPC-scale in terms of carbon footprint per occupant. [14].

The ranges laid out are based on personal local observations and experiences with final year Master students of Green and Energy Efficient Buildings (MGEEB) who have been given the task to measure their daily and weekly energy consumption in their households. All of the students which were quite senior lived with their own families. A household with less than 100 kg/occupant was considered low "green", whereas >210 was gauged as quite high "red".

The following table in Figure 7. reveals when elaborating on house No. 3 and 5 that even the moderate use of an air conditioner will be the trigger for electricity costs of about four times the price of a house cooled just by mechanical fans (house 1, 2 and 4).

The TEPC appeared to be applicable for houses which are already in use, and it might be applicable for future houses. It has been found that examining current households can serve as a great benchmark and guide to avoid the generation of CO_2. These are the average scores during three times daily measurements with "typical" overcast *i.e.*, sunny/cloudy weather conditions, and two consecutive days of no rain. As an accuracy option for future R&D it can be noted that at least 10 afternoons during the heat peak hours from 5–8 pm with different weather conditions should be considered).

By the scales, the students learnt best practice from each other to avoid the permanent, unnecessary production of stand-by hot water, sensors and the use of thermostats to interrupt the flow of electricity in case of older generations of non-inverted air conditioners.

Description	Built in (Year)	Built-up m²	Additional Non-Standard Appliances	Costs/ KWh p.m. => per m²	kg CO2 per capita/ p.m.
1.Flat 5 storey	1986	96 m²	-	8 EUR = 160 kwh => 0.08 EUR per m²	1 person 160/1= 160 kg
2. 3 storey Terrace House	2008	67 m²	-	15 EUR = 292 kWh => 0.22 EUR per m²	2+1 persons 253/2.5= 121 kg
3. 2 storey Terrace House	1997	98+88 Ground floor = 98 m² First floor = 88 m²	Water Heater AC	37 EUR = 700 KWh => 0.19 EUR per m²	3 persons 607/3= 202 kg
4. Taman Semi-Detached	2009	83 m²	-	9 EUR = 37 kWh => 0.11 EUR per m²	2+1 persons 138/2.5= 101 kg
5. Low-Cost Highrise (level 10, 25 storeys)	?	65 m²	Water Heater	41 EUR = 775 KWh ⇒0.63 EUR per m²	6 persons +1 baby 672/6= 139 kg

Figure 7. Malaysian case studies on carbon footprint per m^2 (column 6, costs in EUR) and per occupant (in kg CO_2p.a. in EUR) [14].

3.3. Thermal Comfort/Thermal Sensation

We define "thermal comfort" as the state of mind that expresses satisfaction with the temperature, humidity and velocity of the surrounding environment (according to the ISO 7730 or, likewise [17,18]. Together with (a) environmental sustainability and (b) long-term cost saving; (c) creating and maintaining thermal comfort for occupants of buildings or other enclosures is the third of the three important output objectives of triple green building architecture and engineering. Thermal comfort constitutes a basic individual need. Presuming it is granted, in a concerted effort with other physical needs, it enables us to climb up the ladder of Maslow's renowned motivational pyramid of needs from basic needs. Conversely, in its absence, any thermal gain or loss above or beyond its essential borderlines may generate a sensation of discomfort. Examples are school children in tropical low-cost buildings who are expected to do their homework in temperatures of 32 °C and above. They become victimised as they will not be in a condition to perform well.

3.3.1. Ambient Temperature

A typical Northern conception of thermal comfort has been based on the idea that the inside temperature for offices should be 21.1 °C on average with variations of ±2.5 °C (like in the ASHRAE Handbook [17]. Of course, in a cold country every °C that does not need to be heated during the predominant cold seasons can save tremendously on energy and budget. In recent years, the average figure for thermal comfort has been even proposed to be altered for European offices to 24.5 °C, which means an enormous deviation from the internationally renowned previous ASHRAE-standard. For Northern countries, this is questionable for environmental reasons because of higher energy demand. For tropical countries, however, it sounds very reasonable. Busch [19] carried out a pioneering field study for Thai offices in Bangkok and found that the optimum thermal sensation for air conditioned and naturally ventilated buildings alike was between 24.5 °C and 28.5 °C. A similar range of "neutral" conducive temperature was determined for a Malaysian School (Ibrahim Hussein, M. Hazrin A. Rahman [20], based on PMV regression of 25.9 °C with a comfort range between 24.4 °C and 27.4 °C. The trend to increase the temperature in offices and public cooled down areas also follows the in-part demise of the common dress code with suits and ties. This trend was translated into the 2011 policy e.g., by the Malaysian government requesting all state-owned office buildings to set-point the temperature not lower than 24 °C.

Abdul Rahman [21] in his ground-breaking study found that the most comfortable indoor temperature in Malaysia for tropical residential areas ranges higher than as for offices. His tolerated set-point of 25.5–28 °C narrows down the general recommendation by the World Health Organization (1990) ranging from 18–28 °C to its upper limits. Similarly, a research team from the University of Technology Malaysia concluded in their research that a 2.5 °C range between 26.1 °C and 28.6 °C is optimal in tropical countries, even for adapted people from Northern countries [14]. Others' and our own findings clearly confirm that the optimal residential area temperature for a normal distribution including about 85% of tropical occupants at its highest comfortable end should not be lower than 23.6 °C and not exceed 28.6 °C. As a conclusion, "the comfort band for the Kuala Lumpur area for all building types is between 23.6 °C and 28.6 °C with an optimum medium temperature in Malaysian households of 26.1 °C" with the upper space limit (USL) set at 28.6 °C [14]. Two reasons can be given. (1) The lower cost when putting the highest set-point in a tropical warm country; (2) the perception by people living in tropical regions is different from those in temperate and cold regions (Wang and Wong [22] or Singh *et al.* [23]. The perception is based on lifestyle and habits, and economic necessities. All of these factors contribute to understanding the following comparison.

As researchers proved several decades ago, the acceptable high temperature of 28.6 °C is not necessarily only restricted to Caucasians or any other human from the colder hemisphere who accepts lower temperatures, or a tropical human being who favours higher temperatures.

Devising the tropically adapted concept of energy performance for thermal comfort with these higher temperature bandings can cause a steep increment in terms of energy saving potentials. A saving of 4%–7% of CO_2 and energy costs with each degree centigrade the temperature is increased can be achieved [24].

Unfortunately, with the USL (upper space limit)temperature set to its highest end at 28.6 °C, in a typical uninsulated concrete building, walls, windows, ceiling and roof still serve inadvertently as permanent heat traps. Hence, tropical thermal comfort without the aid of strongly blowing ventilators or air conditioners can often not be achieved. During the heat peak hours of a typical overcast sunny/cloudy day, this is difficult. Even in kampong (Malay village) areas located outside the city, which are affected by the heat island effect [4], the ambient temperature during the heat peak hours often exceeds an intolerable 30 °C. Table 1 zooms in different temperature comfort bandings.

Table 1. Comparison of different thermal comfort definitions [25].

Country	Min. Temp. °C	Median Temp. °C	Max. Temp. °C	Reference
Northern Countries	19.1	21.1	23.1	[17]
"New" Approach" (offices)	22.5	24.5	26.5	[26]
Tropical Malaysia (residential)	23.6	26.6	28.6	[9]

Apart from environmental issues, it can be concluded that whether a building yields blue, dark green or light green in terms of thermal comfort temperatures is a matter of individual sensation and well-being, verified by normal distribution based studies with different populations.

In a DoE setup (design of experiments), the average scores are now measured three times daily during the dry season's heat peak hours with sunny/cloudy conditions after two consecutive days without rain. As indicated in Figure 8, thermal comfort in Northern and Tropical zones may have different weightages. If the temperature is between 26.8 and 28.6 °C, the highest scores can be achieved for the tropics. The definition of green in Northern countries constitutes lower temperatures, receiving the highest scores in the Northern atmosphere. However, the longer a person lives in a tropical country, the more he or she seems to be adaptable to the local standard of thermal comfort [27]. Altogehter, the overall bandings of ambient temperatures provided by the TEPC become clear by Figure 8:

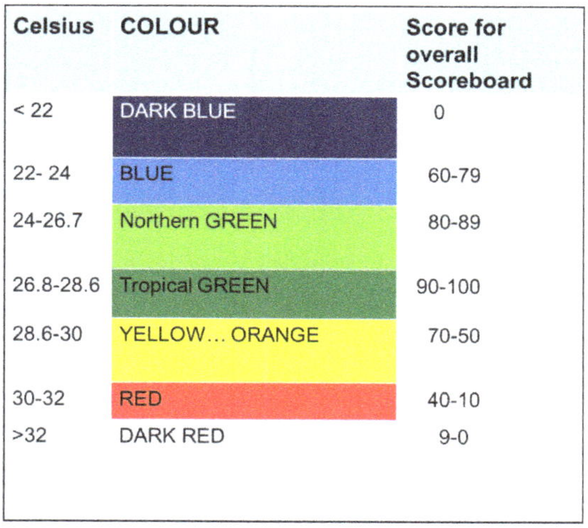

Figure 8. Thermal comfort: residential ambient temperature banding [14].

3.3.2. Relative Humidity (RH)

So far, as the second parameter of thermal (dis)comfort, relative humidity (RH) has not yet been considered for the TEPC. At this point in time, the author removed relative humidity as a consideration, but not just for simplicity reasons. The degree of moisture in the air is volatile and admittedly still hard to control by water- or velocity-based green cooling tools without the aid of air conditioners. Conversely, green cooling by water or air may make the already existing humidity more humid. Only if we would be able to live according to rules of a green lifestyle, we would not consider the high humidity as a serious or even the prime issue when it comes to thermal comfort banding; the tool is not yet complete. Subsequently, the relative humidity's effects for thermal sensation are determined by different metabolism rates, our behaviour of taking showers, not eating too late and too heavy, and by utilising stronger ventilation to decrease the ambient temperature to diminish the effects of high humidity at the same time.

After some years, researchers have abandoned the idea of accepting the cold-hemisphere-borne ASHRAE-standard in terms of temperature; it is the time to question the same standard in the tropical experience having an obsolete borderline of 60% for relative humidity. Questioning this standard, the author does not stand alone. Local tropical researchers confirm have confirmed in conferences that "above 60%" is a dissatisfactory borderline when, across the board, the RH in urban and sub-urban areas is almost always 10%–20% higher. ASHRAE is only statically defined by the amount of moisture that can be digested by our skin without causing sweating. The 60% standard presumes that tropical countries without artificial dehumidification will never have livable homes—unless they use air conditioners or not yet marketable dehumidifying gadgets. However, as studies have not yet been undertaken, there is nothing in writing to verify the question behind the 60% borderline. Hence, we must await further applied research in the tropics that comes up with more empirical evidence in order to argue this issue in depth beyond academic discussion.

3.4. Cost Considerations

The last variable is dedicated to answering the question of the 3rd angle of the magic triple green triangle: Who can afford our green and energy efficient building? Whereas the other three criteria above tend to refer more to static measurement methods, the standards of cost considerations must be adapted, because the individual household's purchasing power via monthly cash flow p.m. is considered along the rainbow colours. Hence, this is a vital subjective question, as it is widely perceived that green buildings are necessarily quite expensive and therefore a NO-GO for mid- and low-income households. Therefore, those who know and would appreciate acquiring a greener home keep believing that they are not able to purchase a new unit. Those who are already inhabitants refrain from retrofitting their existing one due to cost considerations.

For new buildings we can state that by following the European principle a passive house (as the most radical version of the green and energy efficient building) may not exceed 110% of the investment costs of a conventional red building. Due to cost savings from the first moment the building is in operation, reasonable pay-back periods of 1–5 years can be expected [28]. In terms of the following scale echoing the values for a new residential building, this is still considered light green.

RM	COLOUR	Score for Overall Scoreboard
Cheaper than standard house	DARK GREEN	100
100 < 110%	LIGHT GREEN	90
111 < 120%	YELLOW	70
121 < 130%	Light ORANGE	50
131 < 140%	Dark ORANGE	30
140 < 150%	RED	15
> 150%	DARK RED	0

Figure 9. Exemplified red–green TEPC-scale in terms of affordability (new buildings). [14].

The case of an existing standard residential building to be retrofitted is even a more subjective question and answer. Whether it can be greened like in the case of the reference buildings with an additional investment of less than 200 EUR and the possibility to break even after a maximum 5 years, it might be "green" or "red". This will mainly depend on three parameters: (a) the household's income; (b) the monthly cash flow and, first and foremost; (c) the readiness to invest in green. Furthermore, one caveat and prerequisite is that the building does a "smart" job. That means green cooling plus building automation can adjust the desired temperature according to the occupant's wishes at heat peak hours, and react flexibly due to the building frame's ability to serve for already much colder temperatures compared to a standard building (elaborated above under pillar 2—electricity consumption).

Hence, the following scorecard is more subjective and situational. It presumes that this mid-class household in a developing tropical country has a monthly cash flow of 600 EUR. Therefore, it could easily digest green retrofitting expenditures of 12 times 200 EUR p.a. and would still have another 100 EUR for other cost positions. That would mean that 200 EUR could easily be absorbed, and up to 400 EUR are still at "light-green" status if the household is ready to invest in green, as elaborated further on in Figure 10.

EUR	COLOUR	Score for Overall Scoreboard
< 200	Dark GREEN	100
< 400	Light GREEN	90
< 600	YELLOW	70
< 800	Light ORANGE	50
< 1,000	Dark ORANGE	30
< 1,200	RED	15
> 1,400	DARK RED	0

Figure 10. Exemplified red–green TEPC-scale in terms of affordability (existing buildings) [14].

In the dark red end of the spectrum, this exemplified scale is left open on purpose with an amount of "above 1400 EUR" which might be quite hard to invest for a middle class income household. Of course, the "below 200 EUR" rule for dark green needs to be adapted not only towards the financial capacities of the occupants which are usually the owners, but also towards the size of the building.

Both scales (new and retrofitting) can now be compared with the payback periods of the necessary green investments. The ROI-calculation will be made due to lower operational costs in terms of energy consumption.

4. Case Study

Within the limited scope of this paper and with the infancy of the TEPC, it is not yet possible to demonstrate the practical value of the TEPC through a significant number of cases. Among those that had been scrutinised by the author in detail are three out of the four common tropical residential houses, *i.e.*, bungalow, terrace house, village house and high rise apartment [15]. In order to illuminate the applicability of the TEPC, prior to generic research the five examples in Figure 7 above for retrofitting have been referred to for the criterion CO2consumptiononly. Out of these, due to the limited space, only case 3 as a terrace house has been chosen to demonstrate how the tool works and to demonstrate the operability of the TEPC for a tropical house (Figures 11 and 12).

- Typical Terrace House Center (Kuala Lumpur Area (built 1997)
- Size: Ground floor = 98 m^2, first floor = 88 m^2 (cooling load 558 m^3), Car park area = 56 m^2
- Location: Suburban area (not affected by city heat stack effect) with evening temperatures at peak heat days reaching tropical thermal comfort level at sunny/cloudy conditions between 8 and 12 p.m.
- Occupants: 5 Adults
- The average electricity consumption p.m. is about 700 kWh = 46$ per month. The following appliances are in use (one common air conditioner used 8 h usually during nighttime)

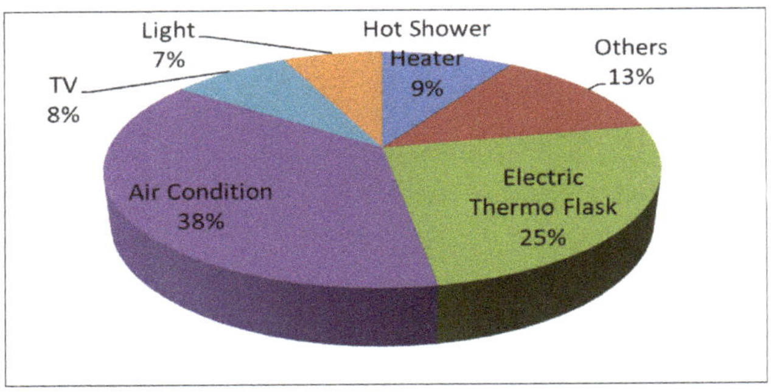

Figure 11. Energy consumption of terrace house Kuala Lumpur [14].

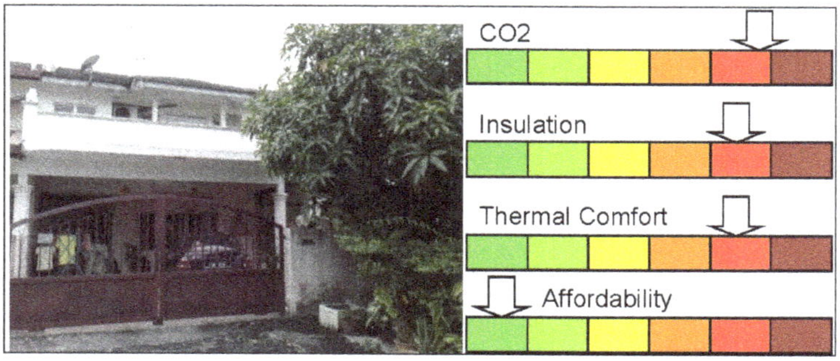

Figure 12. Case study oftypical terrace house center (Kuala Lumpur) area (1997) and applied TEPC) [14].

As indicated by the arrows in Figure 8, the TEPC will yield the following results:

(1) CO_2 = Due to the air conditioner and the all-time stand-by thermo flask, the electricity consumption can be classified as red.
(2) Building envelope's radiant temperature average (measured as proposed above at overcast stable weather heat peak hours at 4.30 p.m.): 32 °C = red.
(3) Thermal comfort.

At the same peak hour around 4.30 p.m., the thermo-logger yielded 31 °C = red. However, the lowest temperature during the nighttime is 23 °C = blue (night time average temperature sleeping room at the same day with air conditioning unit)

(4) *Affordability* = green:

In order to bring items 2 and 3 into the green zone, the following investments can be proposed:

(1) Air conditioning unit in three sleeping rooms to be fitted with interrupter, auto-switch into cross-ventilation mode. Lighting can be replaced with low-energy or LED-tubes r bulbs) (300 EUR).
(2) Insulation of the roof (3000 EUR).
(3) Add on insulation foldable wood-wool shutters to all eight windows (100 EUR × 8 = 800 EUR).
(4) Insulation of walls and doors (6000 EUR)

As the cash flow of this family with three adults working equals 900 EUR per month, this expenditure can be considered as affordable green.

As a conclusion, even though this building is in RED areas in terms of CO_2 emission, insulation and thermal comfort if no air conditioning is used, it has very high potential compared to other case studies. The total family income and the monthly cash flow are considerably high. Hence, the chances to green this building are equally high if the occupant (who is also the landlord) creates awareness and applies the EPC to guide the family.

5. Conclusions and Outlook

In order to assist tropical countries to reduce their CO_2 emissions, the TEPC can play a role as it h continually done in some European counties and a growing number of other countries. It can become a widely popular tool as it contains not only an awareness model for thermal comfort, but also a clear, affordable business model for landlords and even tenants. Research in mock-up buildings, simulations and negotiations with internationally operating certification bodies like the TÜV or the Green Mark® tool will demonstrate in the near future whether the approach is viable for developing countries in the tropical and subtropical belt. However, these are potential impasses we may have to face.

- Lack of reliable research data. So far, we only can rely on selected case studies rather than on a multiple of generic cases to green tropical sustainable and energy-efficient buildings.
- Availability of system knowledge (playing together as a "green orchestra"): the TEPC has yet to explore the idea of looking at the building from a holistic perspective (especially the backlog in terms of factoring in life cycle analysis and humidity issues).
- Transparency of professional research-based benefits (Return of Investment -ROI). (Reliable data have not yet been generated to forecast the monthly saving on the part of the electricity bill to establish impetus to green a building further.
- Availability of resources: A visit to green expos shows that both knowledge and tropical greening tools are available. Nevertheless, the market is not transparent when it comes down to sourcing green suppliers that can provide solutions.
- No financial incentives (to create market value like with some European energy regulations). Success will be achievable only if policy makers take the first step by the timely and staggered introduction of energy regulations to enforce green strategies via the TEPC.
- Mindset of owners and tenants (short-term thinking and other priorities—Demand Preference Structure (DPS)). Of course, the greatest challenge of all is to create a change of mindset within the population. Beyond academia, this can only be achieved with a generic roll-out plan compiled by a committed team spearheaded by a highly effective business organisation.
- The extension to humidity measurement, along with other potential effectors, e.g., life cycle analysis, waste reduction, recycling green building materials, utilisation of renewable green energy or other indicators is possible. However, there is a danger that it could overrule and even undermine the tool's Unique Selling Proposition (USP) of the TEPC compared to more promising, but at the same time, much more expensive indices available on the green building market.

Conflicts of Interest

The author declares there is no conflict of interest.

References

1. Najib, S.; Rahman, A. Speech as Prime Minister of Malaysia. In Proceedings of the COP Conference Copenhagen Climate Change Conference, Copenhagen, Denmark, 7–19 December 2009.
2. Byrd, H. *Energy and Ecology. A View of Malaysia Beyond 2020*; USM Publication: Penang, Malaysia, 2008.
3. DeNA. *Erneuerbare-Energien-Wärmegesetz und Energieeinspar-Verordnung anwenden (Briefing and Practical Dialogue: Renewable-Energy-Law and Application of Energy Saving Regulation)*; DeNA: Berlin, Germany, 2010. (In German)
4. Ahmad, S.H.; Mahyuddin, R. Natural Ventilation of Indoor Air Temperature: A Case Study of the Traditional Malay House in Penang. *Am. J. Eng. Appl. Sci.* **2010**, *3*, 521–528.
5. Luciana, L. *Green Building Guide*; Malaysian Italian Chamber of Commerce (MICC): Kuala Lumpur, Malaysia, 2011.
6. Akanle, T.; Appleton, A.; Kulovesi, K.; Schulz, A.; Sommerville, M.; Spence, C.; Yamineva, Y. Summary of the Copenhagen Climate Change Conference. Available online: http://www.iisd.ca/vol12/enb12459e.html (accessed on 30 May 2014).
7. Weglage, A.; Gramlich, T.; von Pauls, B.; von Pauls, S. *Energieausweis–Das große Kompendium: Grundlagen–Erstellung–Haftung (Energy Performance Certificate–Basics–Procedure–Legal Issues*; Springer Vieweg Verlag: Stuttgart, Germany, 2009.
8. Building Sector Energy Efficiency Project (BSEEP), Public Works Department Malaysia. In Proceedings of the Prime Minister's Green Technology and Climate Change Council Meeting Realizing the 3 Key Benefits of Energy Efficient Buildings, Kuala Lumpur, Malaysia, 9 January 2014.
9. Sabarina, S.; Nor Zaini, I.Z.; Mohammad, S.M.; Mohd Ghadaffi, S. *Achieving Thermal Comfort in Malaysian Building: Bioclimatic Housing*; Universiti Teknologi: Skudai, Malaysia, 2007.
10. SEDA. Sustainable Energy Development Authority Malaysia. Available online: http://seda.gov.my/overview_of_fit_concept_in_malaysia.html (accessed on 17 November 2014).
11. Zalina, S. *Building Performance Assessment Systems: The Emergence and the Variety*; Universiti Putra Malaysia (UPM) Housing Research Centre: Serdang, Malaysia, 2012.
12. German Federal Ministry of the Economy, Energieeinsparung (Energy Saving). Available online: http://www.bmwi.de/BMWi/Navigation/Energie/Energieeffizienz-und-Energieeinsparung/energieeinsparung.html (accessed on 30 May 2014).
13. Maul Green Building Consultancy. Available online: http://www.energieberatung-maul.de (accessed on 16 November 2014).
14. Abdul Malek, A.R.; Muna Halim, A.S.; Azizi, B.; Mohd Rodzi, I. *Towards a Low-Energy Building Design for Tropical Malaysia*; Universiti Sains Malaysia Press: Penang, Malaysia, 2009.
15. Wagner, K. *The Making of a Green and Energy Efficient Tropical Residential Building*; Universiti Kuala Lumpur Press: Kuala Lumpur, Malaysia, 2014.
16. Pan, Y.; Huang, Z.; Wu, G. Calibrated building energy simulation and its application in a high-rise commercial building in Shanghai. *Energy Build.* **2007**, *39*, 651–657.

17. ASHRAE. *Thermal Comfort. Fundamentals Volume of the ASHRAE Handbook*; ASHRAE: Atlanta, GA, USA, 2008.
18. Djongyang, N.; Tchinda, R.; Njomo, D. Thermal comfort: A review paper. *Renew. Sustain. Energy Rev.* **2010**, *14*, 2626–2640.
19. Busch, J.F. Thermal Responses to the Thai Office Environment. *ASHRAE Trans.* **1990**, *96*, 859–872.
20. Ibrahim, H.; Rahman, M.H.A. Field Study on Thermal Comfort in Malaysia. *Eur. J. Sci. Res.* **2009**, *37*, 134–152.
21. Abdul Rahman, A.M. Housing design in relation to environmental comfort. *Build. Res. Inform.* **1995**, *23*, 49–54.
22. Wang, L.; Wong, N.H. Applying Natural Ventilation for Thermal Comfort in Residential Buildings in Singapore. *Architect. Sci. Rev.* **2007**, *50*, 224–233.
23. Singh, M.K.; Mahapatra, S.; Atreya, S.K. Thermal performance study and evaluation of comfort temperatures in vernacular buildings of North-East India. *Build. Environ.* **2009**, *45*, 320–329.
24. The Star Newspaper, Green Efforts Start at 24 °C. *The Star*, 12 August 2011, 2.
25. Wagner, K. Tropical Thermal Comfort and Adapted Tropical Green Residential Housing. In Proceedings of the Conference for Affordable Quality Housing, Putrajaya, Malaysia, 13 March 2013.
26. Braatz, W.; Suparoek, J. *Thermal Comfort. Unpublished Manuscript*; Rosenheim, Germany, 2008.
27. Ellis, F.P. Thermal Comfort in Warm, Humid Atmosphere: Observations in a Warships in the Tropics. *J. Hyg.* **1952**, *50*, 415–432.
28. DeNA. *Renewable Energies Made in Germany*; DeNA: Berlin, Germany, 2009.

A Method for Proposing Valued-Adding Attributes in Customized Housing

Cynthia S. Hentschke, Carlos T. Formoso, Cecília G. Rocha and Marcia E. S. Echeveste

Abstract: In most emerging economies, there has been many incentives and high availability of funding for low-cost housing projects. This has encouraged product standardization and the application of mass production ideas, based on the assumption that this is the most effective strategy for reducing costs. However, the delivery of highly standardized housing units to customers with different needs, without considering their lifestyle and perception of value, often results in inadequate products. Mass customization has been pointed out as an effective strategy to improve value generation in low-cost housing projects, and to avoid waste caused by renovations done in dwellings soon after occupancy. However, one of the main challenges for the implementation of mass customization is the definition of a set of relevant options based on users' perceived value. The aim of this paper is to propose a method for defining value adding attributes in customized housing projects, which can support decision-making in product development. The means-end chain theory was used as theoretical framework to connect product attributes and costumers' values, through the application of the laddering technique. The method was tested in two house-building projects delivered by a company from Brazil. The main contribution of this method is to indicate the customization units that are most important for users along with the explanation of why those units are the most relevant ones.

Reprinted from *Sustainability*. Cite as: Hentschke, C.; Formoso, C.; Rocha, C.; Echeveste, M. A Method for Proposing Valued-Adding Attributes in Customized Housing. *Sustainability* **2014**, *6*, 9244-9267.

1. Introduction

There has been increasing diversity in household profiles in the last few decades, due to major changes in contemporary lifestyles. Consequently, dwellers are no longer satisfied with standardized houses, and there is a growing demand for housing products that are able to meet individual needs [1]. Indeed, previous studies on the housing market of different countries, such as Brazil [2,3], México [4], and U.K. [5], indicate that standard products often lead to dwellers dissatisfaction. In this context, the application of the Mass Customization (MC) strategy can potentially increase the value of housing, by delivering products that fulfill specific requirements of different customers through flexible process and organizational structures with delivery times and costs similar to mass production [6–9]. This strategy has been successfully applied by some house-building companies from different countries, including U.K. [5,10], Netherlands [11], México [4], Brazil [12,13], and especially Japan [14,15].

In mass customization, organizations must identify attributes and options that can effectively generate value for clients [9,16]. One of the key decisions in MC is the level of customization of the product, which is determined by the range of choices that are offered to clients [9]. On one hand, the boundaries between MC and craft production lies in limiting the range of choices so that efficiency in production can still be attained [17]. On the other hand, clients' perceived values must be considered for defining the range of options to be offered, since not all options might add value to clients [17,18].

However, not much has been published on how to determine the level of customization based on clients' requirements and organizational capabilities [9,12,19]. In fact, only a few studies have focused on the definition of attributes and options to be offered [19]. In addition, those studies are mostly focused on demand forecast and suggestions based on clients' prior preferences, instead of market segmentation [19]. As a result, defining customizable attributes and options from a value generation perspective remains largely unexplored. This also applies to the housing sector, since house-building companies often define customizable options without a systematic analysis of clients' profile and requirements [12,20].

This paper aims to address this gap by proposing a method to define value-adding attributes in customized housing projects, which can support decision-making regarding product attributes to be customized and the range of options to be offered. This method builds upon the means-end chain (MEC) conceptual model developed by Gutman [21]. This model organizes product attributes, consequences from product use, and clients' values in a hierarchical structure, normally represented by a tree diagram, named hierarchical value map (HVM) [22,23]. This map is a summary of information about value generation that might be used as a visual device for supporting decision-making [22].

The MEC conceptual model was chosen due to some advantages in relation to some currently used tools and practices, such as Quality Function Deployment (QFD), and post occupancy evaluation surveys: (i) it is built upon an in-depth, well structured conceptualization of value [21,24]; (ii) it provides an understanding of value generation by explicitly exploring the consequences of products-in-use and the generation of benefits from the point of view of the final customers [21,24]; (iii) there are a number of data collection and processing techniques, known as laddering, that can be effectively used for generating hierarchical value maps [25–27]; and (iv) hierarchical value maps can be refined and be used as visual devices to support decision making [22,28].

The proposed method was tested by applying it in the evaluation of two construction projects delivered by a small-sized construction company, based in the Metropolitan Region of Porto Alegre, Southern Brazil.

Besides improving value generation in house building projects, the development of the method aims to make a contribution in terms of making those types of projects environmentally and socially sustainable. Providing housing tailored to clients specific requirements' can potentially reduce waste resulting from the changes carried out by dwellers after moving in. In addition, such changes often compromise air and lighting conditions [20], thus eventually leading to increases in energy consumption. Finally, self-expression and personal identity have been pointed out as important qualities of dwellings [16]. As a result, providing tailored housing can increase dweller sense of identity and ownership, thus improving care and maintenance and potentially expanding the dwelling life cycle.

2. Theoretical Framework

2.1. Defining the Scope of Customization in House Building

Rocha [12] proposed a conceptual framework for devising customization strategies in the house-building sector, based on the adaptation of a set of concepts related to MC to the specific context of house-building. Three core categories of decision were proposed in this framework for defining the scope of customization: customization units, solution spaces, and classes of items. This set of decisions aims to define the product variants that will be offered to clients [12].

"Customization units" are the building blocks of a customization strategy: they consist of customizable attributes of a product and the range of options to be offered to clients [12]. Once customization units are defined, the processes for delivering each one of them must be defined [9], as well as the visualization approach to be used, *i.e.*, how the customization units are presented to clients and internally to the organization [12]. A key decision in the definition of the customization strategy is the order penetration point, in which the client order first enters the supply chain, differentiating the activities that are demand driven from the ones that are forecast driven [29].

The "solution spaces" decision category defines how the customization units are combined or grouped. This decision category establishes the boundaries within which the customization units can vary, considering limitations in production capabilities [17]. For each solution space, there might be a platform, as well as an order penetration point. A platform is the product part that remains unchanged throughout the different configurations of a product within a solution space [30]. It is necessary to map the customization units and how these are combined so that the solution space can be clearly established in the customization strategy [12].

The "classes of items" category expresses the different properties of the customization units based on the items contained in each one of them [12]. Customization units can include a limited number of items for a given attribute (e.g., three options for color). Yet, as discussed in Rocha [12], they can also be devised for clients to define any item for the attribute under consideration (e.g., specify any color). Finally, customization units might require the selection of an option or allow clients to refuse these options (e.g., no coloring).

The definition of the scope of customization must consider constraints in the production system, so that economies of scope can be achieved, *i.e.*, reductions in the average cost by producing a limited range of products that require similar operations [17]. Therefore, there should be a balance between product flexibility, defined according to customers' preferences, and the performance of the production system [9,18]. It means that it is necessary to understand the trade-offs between benefits provided to customers by customizing products and sacrifices such as increased cost or delivery time [6,17]. One of the key aspects for reaching success in the implementation of MC is the integration between different product development areas, such as design, marketing, production management, and supply chain management, since different types of decisions are involved in the definition and in the implementation of a customization strategy [6,9].

2.2. The Means-End Chain Model and the Laddering Technique

The means-end chain model was originally developed by Gutman [21] to describe how consumers categorize information about products in the memory, seeking to understand their purchasing choices. Subsequently, Woodruff and Gardial [24] expanded this conceptual model, considering not only the desired value upon purchase, but also the value received during the use of a product.

The means-chain model establishes an association between the characteristics of a product-in-use and the motivations that lead to consumption. According to Gutman [21], the means are represented by the products or services in which individuals engage, and the ends are valued states of being. According to Reynolds and Olson [31], the means-end chain model connects the concrete attributes of a product (tangible attributes) with the emotional and personal values (abstract and intangible objectives) through a chain with six levels of abstraction: concrete attributes, abstract attributes, functional consequences, psychological consequences, instrumental values and terminal values.

The concrete attributes are inherent to the product or service, and can be directly perceived (for example, color, brand name). Abstract attributes are aspects of the product or service that cannot be measured or perceived through the senses, such as quality or reputation [23]. The functional consequences emerge from the direct relationship between the consumption of a product and the individual and are related to the utility of the product in a specific use situation [32]. The psychosocial consequences are related to the ability of a product or service to satisfy intrinsic objectives that are symbolic, self-oriented or other-oriented, *i.e.*, projecting an image that is congruent with the norms of meanings of others [32]. The instrumental values are intangible goals related to the behavioral means used to achieve the ends [33]. Finally, the terminal values refer to desired end states (for example, happiness, security, accomplishment) [21,33].

The representation of these hierarchical levels through a visual device, such as a hierarchical value map, can help decision makers involved in the provision of a product to understand the value generation for the users of that product [22]. There are several techniques that can be used for mapping the relationships between the product's attributes, its consequences in use, and desired values, such as association pattern technique (APT), hard laddering [25], and soft laddering [25–27]. In this study, the soft laddering technique was chosen, due to its ability to cope with complex products, such as housing. This technique provides a rich picture of customers' perceived value that can be used for generating new ideas.

The main source of evidence used in the soft laddering technique is a one-to-one interview, consisting mainly of questions like "why is this important to you?" [26]. The term ladder expresses the main goal of this technique: to go from the more superficial to the deeper levels, from the more tangible to the more abstract levels by using open questions. The main outcome is a hierarchical structure for the attributes, consequences and values of a specific product in use. Reynolds and Gutman [26] proposed five steps for the application of the soft laddering technique: (i) in-depth interviews; (ii) analysis of the content of the ladders, which is a sequence of answers by an interviewee to questions ranging from the attribute to the highest level of abstraction, and the codification of those answers in means-end chains; (iii) elaboration of the implication matrix, where the relations between the elements in the means-end chains are analyzed; (iv) construction of the hierarchical value map (HVM); and (v) interpretation.

3. Research Method

Design science research was the methodological approach adopted in this study. It is a form of scientific knowledge production that involves the development of innovative constructions, intended to solve problems faced in the real world, and simultaneously makes a kind of prescriptive scientific contribution [34]. An important outcome of this type of research is an artifact that solves a domain problem, also known as solution concept, which must be assessed against criteria of value or utility [35]. In this research study the proposed artifact is a method for defining value-adding attributes for customized housing, which can be used to support decision-making in the definition of the scope of customization.

Similar to the research process suggested by Lukka [34] for design science research, this investigation was divided into four main phases: (i) choosing a relevant practical problem and understanding the topic from a theoretical point of view; (ii) devising the artifact; (iii) implementing the method in a practical situation; and (iv) analyzing the scope of applicability and the theoretical contribution of the solution.

Phase 1 started by understanding the context of the product development process of social housing. Initially, information about the housing provision programs in Brazil, especially from the My House My Life (*Minha Casa Minha Vida*) Program, was obtained from different sources: articles that reported previous research studies on low-cost housing, official documents about program procedures and minimum requirements, and a small number of open interviews with technical staff of the National Savings Bank (*Caixa Econômica Federal*), the financial organization that operates the funding for most low-cost housing programs in Brazil. Then, some information of the product development process carried out by the house building company involved in this investigation was obtained in two open interviews with the design manager of that company. The scope of product customization offered by the company was analyzed, based on leaflets, architectural drawings, and choice menus for previous projects. Moreover, the research team had access to an existing database of projects already delivered, which contained relevant information about the demand for customization, including client satisfaction surveys, post-occupancy complaints, and design changes demanded by clients.

The first version of the method was devised in Phase 2. The method was divided in several elements: a data collection protocol, a set of statistical techniques for analyzing the data, and some visual devices for presenting information. The development of the data collection protocol was based on previous studies concerned with the evaluation of low-cost housing in Brazil [13,28].

In Phase 3, the proposed method was tested considering the current context of the house-building company, This company was chosen due to the fact that it had a good reputation on the development and construction of low-cost housing projects, and adopted the strategy of customizing housing units in most projects. The company suggested two projects for implementing the method. These were chosen due to the availability of data about clients, and also because both were financed by housing provision programs that allowed the company to offer customized units. In both projects, the clients had three months to make a decision about the customization units. They were asked which changes they would like to make in the housing unit from a pre-set choice menu. Table 1 presents an overview of the two projects.

Table 1. Brief description of the two house-building projects.

Project	Housing Program	Type of Housing Unit	Gross Floor Area	Number of Housing Units	Sample	Delivery Date
DL	Funding to Support Production	1 and 2- bedroom apartments	55 to 64 m^2	64	32	March 2008
DE	Associative Credit	2 and 3- bedroom apartments	64 to 110 m^2	20	10	May 2009

Altogether, 42 housing units were selected for data collection, using the simple random sampling technique with finite population correction. This sampling technique can be used for reducing slightly the sample size when the population is small. The sampling error of 10% and the level of significance of 90% were adopted. For each unit, a face-to-face interview with one of the dwellers was carried out, and, when possible, direct observations of changes made in the apartments were undertaken.

Finally, in Phase 4, a discussion of the results in a one-hour meeting, using the visual displays produced by the method, was carried out with the participation of the design manager, two architectural designers and the company director. Based on that discussion and also on the analysis of the data collection and processing process, the applicability of the method was assessed.

4. Method for Defining Value-Adding Attributes in Customized Housing Projects

Figure 1 presents an overview of the proposed method for identifying value-adding customization units. The method has been devised for organizations interested in defining or refining the space of solutions to be offered in customized house-building projects. The method is divided into five main steps: (i) characterizing the housing products delivered in past projects; (ii) devising the data collection protocol; (iii) collecting data in a sample of house-building projects; (iv) data processing and analysis, including a number of statistical tests; and (v) producing visual devices to support decision making.

The characterization of the housing products consists of mapping the products being offered to the clients, mostly based on the architectural design, and on resources used to advertise the project, such as leaflets, and company website. Based on that information, a clear definition of the space of solutions offered to clients can be obtained. Some customization units can also be added or removed, based on additional data on customization demands obtained from other sources, such as post occupancy evaluations, and surveys with potential customers.

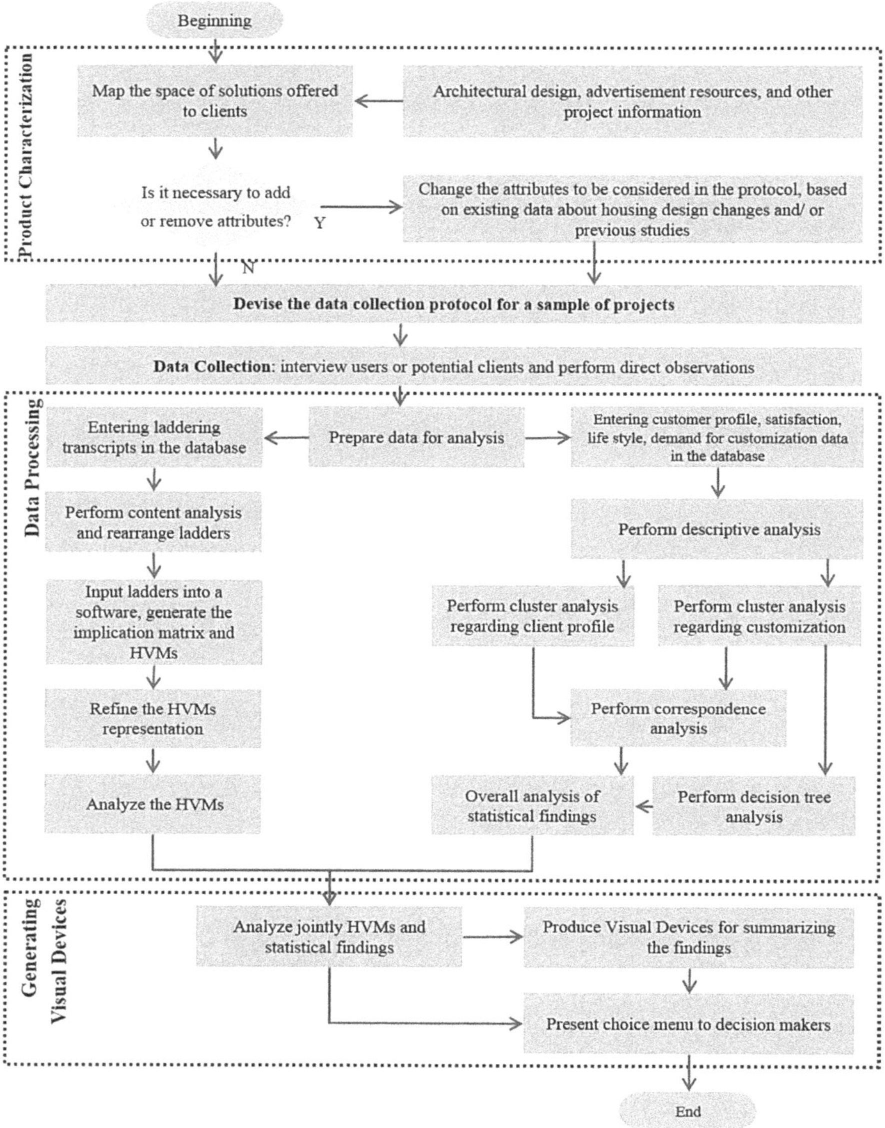

Figure 1. Method for proposing value-adding attributes for customized housing.

In the second and third steps, a data collection protocol is devised and applied in a sample of projects, which might be representative of one or more market segments in which the company operates. This data collection protocol comprises: a semi-structured interview with a sample of users on customer profile, evaluation of satisfaction, and demands for customization; sampling criteria; and guidelines for the direct observation of the housing units, buildings and surrounding areas.

Data processing, the fourth step, is initially divided into two main blocks: (i) the analysis of laddering data, focused on the three most important changes (made or desired) in the housing unit from the point of view of the dwellers; and (ii) statistical analysis of data about the customer profile and life style, demands for customization, and customer satisfaction.

As suggested by Reynolds and Gutman [26], it is necessary to use a content analysis technique to identify a chain of connections between attributes, consequences and values, named ladders. Table 2 presents a typical outcome of the laddering technique, after the codification that results from the content analysis. Reynolds and Gutman [26] suggested that a cut-off point that represents 66% of the relationships should be considered in the map, while Gengler, Klenosky and Mulvey [22] proposed that a balance should be achieved between reduction and retention of information. The cut-off point indicates the minimum number of times a link between two constructs has to be established by subjects in order to appear in the HVM [26]. A software package can be used to support the production of hierarchical value maps—the *LadderUX* software package has been successfully used in this research study.

Regarding statistical analysis, a wide range of techniques can be used. Figure 1 presents the techniques used in this implementation of the method: descriptive analysis, cluster analyses, and correspondence analysis. However, depending on the aims of the evaluation, other techniques could be brought into the method. At the end of this step, an overall analysis of the statistical findings is made.

In the fifth step, a cross analysis of the hierarchical value maps and the results of the statistical analysis should be made. For instance, different hierarchical value maps can be produced for each cluster of dwellers. Finally, a number of visual devices are produced, including different hierarchical value maps and a choice menu, which summarize the data produced by the method.

Table 2. Examples of ladders, codification and reordering.

Raw Data	Codified and Reordered Data
The wall painting didn't give it much life; the white. More about the aesthetics. Much lower quality paint. Feeling good.	ALL ROOMS | WALL FINISHING | QUALITY OF MATERIALS | HYGIENE |
A place to rest. Like a clean and well looked after room. Well-being	AESTHETICS | RESTING AND ENJOYING ONE'S FREE TIME | WELL-BEING

5. Implementation of the Method in an Empirical Study

5.1. Customer Profile

The majority of the families in the two projects were small ones: 90% of them had up to three members, and 36% consisted of couples with children (Figure 2a). However, there was much diversity in the configuration of families, as shown in Figure 2a, confirming the results of the research studies carried out by Formoso, Leite and Miron [3] and Villa [36]. In terms of family income, the large majority of families earned more than the equivalent of three minimum wages (m.w.); most of them earning between six and nine minimum wages (Figure 2b).

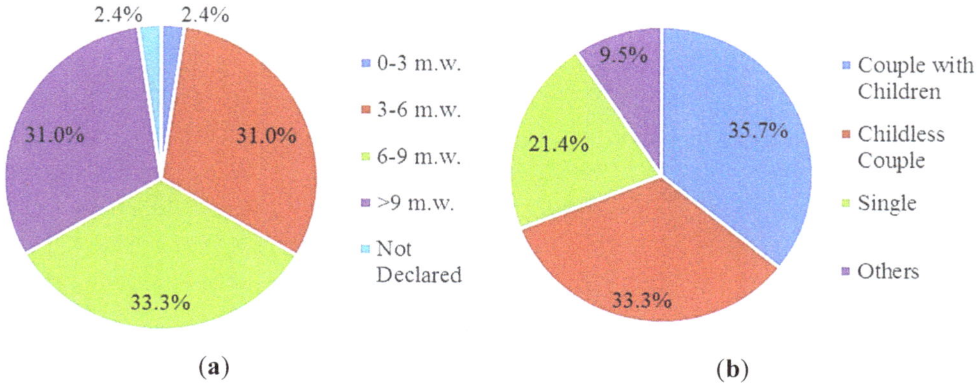

Figure 2. Profile of the families: (**a**) configuration and (**b**) income (minimum wages).

A two-step cluster analysis was performed using the customers profile information, such as family configuration, income, gender of the head of the household and amount of children. Figure 3 compares three customer profile clusters that were identified: (i) couples with children, in which a female is the head of household (18 families); (ii) family in which a male was the head of household—either couples with children or singles (12 families); (iii) childless couples in which a female is the head of household (11 families). Those clusters represent 97.61% of the sample (41 families). An average silhouette measure of cohesion and separation of 0.3 was obtained, indicating that the quality of the cluster solution was fair. In both clusters that had women as heads of household, she predominantly had a university degree. In terms of income, the households headed by men had the lowest income, and the households without children the highest.

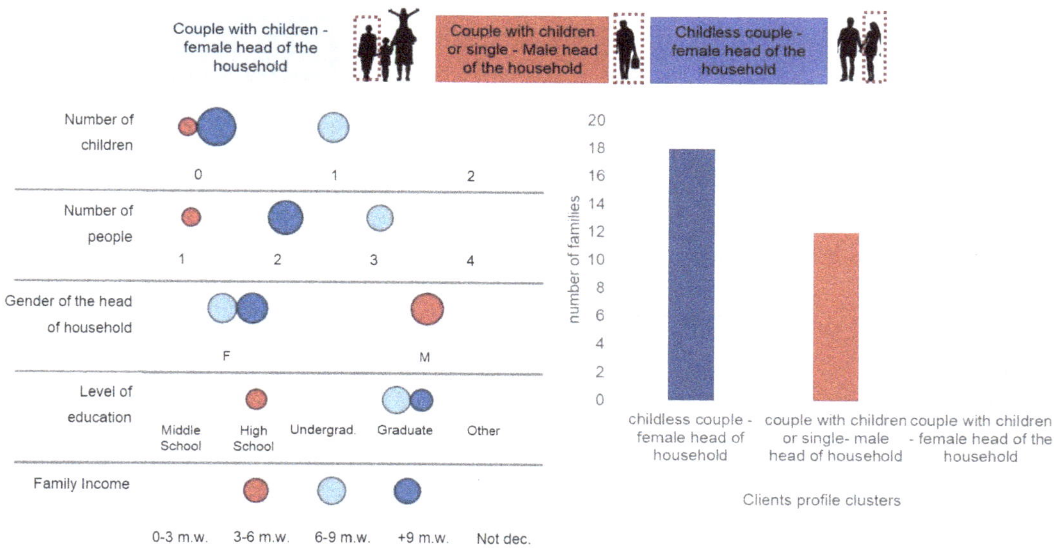

Figure 3. Comparison of customer profile clusters.

5.2. Frequency of Product Changes

For the analysis of the frequency, three types of product changes were considered: (i) asked by the client at the design stage as product specifications and carried out by the company; (ii) made by users after delivery; (iii) still desired by users. Table 3 describes the solution space offered by the house building company, and compares it to the product changes made or desired by users. It is worth noting that for most of the design changes made by the company an extra payment was required. Both Table 3 and Figure 4 indicate that most product changes demanded or performed by users after delivery had been offered by the house building company before sale. This might indicate that some of the clients were not aware of the customization units or classes of items offered by the company. In other cases, users decided to choose an option (e.g., finishings) that was not among the ones offered by the company.

Table 3. Comparison of customization units and options offered by the house building company with product changes made or desired by users.

Customization Units	Design Changes Performed by House Building Company	Product Changes Made or Still Desired by Users
Kitchen and laundry room wall finishings	Standard (ceramic tiles up to half-height), or ceramic tiles up to ceiling	Ceramic tiles up to ceiling, or washable and anti-mold paint
Kitchen and laundry room floor finishings	Standard (Ceramic tiles) or porcelain tiles	Ceramic tiles, or porcelain tiles
Living and dining room floor finishings	Standard (none), ceramic tiles or wood veneer	Ceramic tiles, porcelain tiles or wood veneer floor
Living and dining room wall finishings		Other colors, or washable and anti-mold paint
Bedroom floor finishings	Standard (none), ceramic or wood veneer	Ceramic tiles, porcelain tiles or wood veneer
Bedroom wall finishings		Other colors, or washable and anti-mold paint
Bathroom floor finishings	Standard (Ceramic tiles), or porcelain tiles	Ceramic tiles, or porcelain tiles
Ceiling finishing	Standard (Painted slab), or Gypsum lining on the living and dining room	Anti-mold paint, or gypsum lining in all rooms
Sanitary ware (sink and toilet)	Standard (*colonial* line), *laguna* line and *valentina* line	Sink over a cabinet, other lines
Laundry tub		Remove for installing a washing machine, or change for a new one
Electrical Services	Change voltage, change stove and hood outlets position	Change voltage, change outlet position, change wiring
Cable TV		Installation of cable TV
Air conditioning	Include outlets	Include outlets and installation, and desire for more spots

Table 3. *Cont.*

Customization Units	Design Changes Performed by House Building Company	Product Changes Made or Still Desired by Users
Kitchen and Living and dining room Layout	Standard in DE (Segregated), or Standard in DL, optional in DE (American kitchen)	Desire for an open-plan kitchen
Doors (opening and position)	-	Change the kitchen door for a sliding door or removing it
Building envelop	-	Repair cracks and ineffective window sealing
Balcony	-	Desire for a balcony
Parking space roof	-	Desire to have roof in the parking space
Acoustic insulation for slabs and between housing units walls	-	Desire for acoustic insulation
Other	-	Desire for radiant floor heating, paint doors and change doorknobs, and add a bedroom

As shown in Figure 4, the most frequent changes made or desired in the housing units were concerned with finishings. Regarding the living and dining room, and the bedroom, even though around 40% of the housing units had customized floors at the design stage, approximately 55% of the users performed changes in the wall and floor finishings of those rooms after delivery, and 25% still desire those product changes. The need for those customizations may be related to the lifestyle of users, who tend to privilege social interaction and entertainment in those areas. It is worth noting that some of the product changes were the result of a number of necessary repairs due to building pathologies, such as cracks, ineffective window sealing, low quality of painting, and poor acoustic insulation.

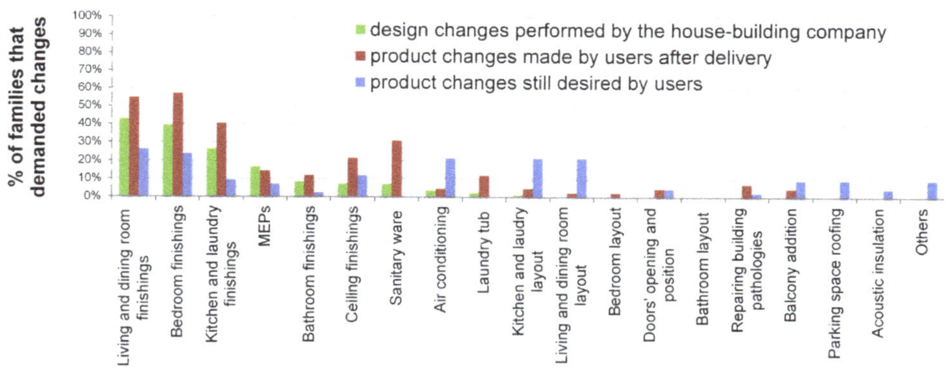

Figure 4. Percentage of families that demanded design changes before construction, or that made or still desire product changes after delivery.

In the case of the laundry and kitchen finishings, all units were delivered with ceramic tiles on the floor and walls up to half-height, which had specifications changed by customer demand in 25% of the units. Forty percent of customizations made after delivery of the unit consisted mostly of painting the walls or completing the tiles up to the ceiling. According to the respondents, their dissatisfaction with those finishings was due to the difficulty to clean and the low quality of the paint. Moreover, 30% of the users changed the sanitary ware after the purchase of the unit, due to aesthetics reasons.

Most of the desired items are related to attributes that the product does not have or are not feasible to provide due to technical reasons. For instance, the installation of air conditioning, desired by 20% of the dwellers was hindered because it would require changes in the façade, and due to the lack of capacity of the electricity grid. In fact, air conditioning had the lowest degree of satisfaction among all attributes due to the difficulty of adding additional equipment. Likewise, the integration of the living/dining room and kitchen, desired by 20% of users, was not possible due to the fact that the kitchen area is too small.

5.3. Clusters of Product Changes

Hierarchical cluster analyses were also performed for product changes with the aim of identifying which customization units were changed together. The association between customization units resulted in sets of clusters (Tables 4–6): five clusters of design changes made by the house building company ($\Phi = 0.492$ and $15 < \delta < 25$); seven clusters of product changes performed by users ($\Phi = 0.337$ and $16 < \delta < 25$); and six clusters of product changes still desired ($\Phi = 0.308$ and $19 < \delta < 25$). The phi coefficient (Φ) measures the association between categorical dichotomous variables, similar to the Pearson correlation coefficient for quantitative variables, and the distance between clusters is measured by the average linkage (δ) [37]. The clusters were ordered from the most to the least frequently undertaken or desired product changes.

Table 4. Clusters of design changes performed by house building company.

Cluster Name/Frequency	Variables
Living and dining room and bedroom finishings 18 families (42.9%)	Living and dining room finishings and Bedroom finishings
Wet areas finishings and sanitary ware 15 families (35.7%)	Kitchen and laundry finishings, Bathroom finishings and Sanitary ware
Air conditioning/9 families (21.4%)	Air conditioning
Laundry tub/7 families (16.7%)	Laundry tub
Ceiling finishings and MEPs/2 families (2.4%)	Ceiling finishings and Mechanical electrical and plumbing installations

Table 5. Clusters of product changes made by users.

Cluster Name/Frequency	Variables
All apartment finishings/26 families (61.9%)	Kitchen and laundry finishings, Living and dining room finishings, Bedroom finishings and Bathroom finishings
Sanitary ware and laundry tub/14 families (33.3%)	Sanitary ware and Laundry tub
Ceiling Finishings/9 families (21.4%)	Ceiling finishings
MEPs and doors/7 families (16.7%)	Mechanical electrical and plumbing installations and Doors' opening and position
Air conditioning and repairs/4 families (9.5%)	Air conditioning and Fixing building pathologies
Layout/2 families (2.4%)	Kitchen layout, Living and dining room layout and Bedroom layout
Others/2 families (2.4%)	Others (floor heating and add a bedroom)

Table 6. Clusters of product changes still desired by users.

Cluster Name/Frequency	Variables
Living and dining room and bedroom finishings and others/15 families (35.7%)	Living and dining room finishings, Bedroom finishings and Others (floor heating, paint doors and change doorknobs)
Living and dining room and kitchen integration and repairs/11 families (26.2%)	Kitchen layout, Living and dining room layout and Fixing building pathologies
Air conditioning and doors/10 families (23.8%)	Air conditioning and Doors' opening and position
Wet areas and ceiling finishings/7 families (16.7%)	Kitchen and laundry finishings, Ceiling Finishings and Bathroom finishings
MEPs and balcony/6 families (14.3%)	Mechanical electrical and plumbing installations and Addition of a balcony
Parking space roofing and acoustic insulation/5 families (11.9%)	Parking space roofing and Acoustic insulation

5.4. Correspondence Analysis

Correspondence analysis was used to associate customer profile clusters and product change clusters, with the aim of identifying the customization units that are more suitable for each customer profile. Figures 5–7 present the results of correspondence analysis for each type of product change. Based on that, some trends became evident:

(a) Households headed by women customize more than those headed by men, and couples without children customize even more, which may also be related to a higher family income or to more time available;
(b) The childless couple cluster, as they have higher income, tend to choose costly customizations, such as floor lining, ceiling finishings, and sanitary ware; and
(c) Families with children apparently invest less in improving housing units, probably because they prioritize other types of expenditure.
(d) There seems to be no relationship between the need for air conditioning points, which stands out as a source of dissatisfaction, and the customer profile;

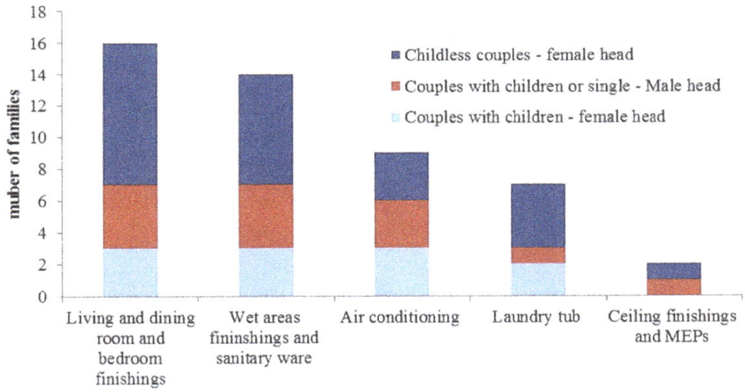

Figure 5. Correspondence analysis of client profile clusters in relation to design changes performed by the company.

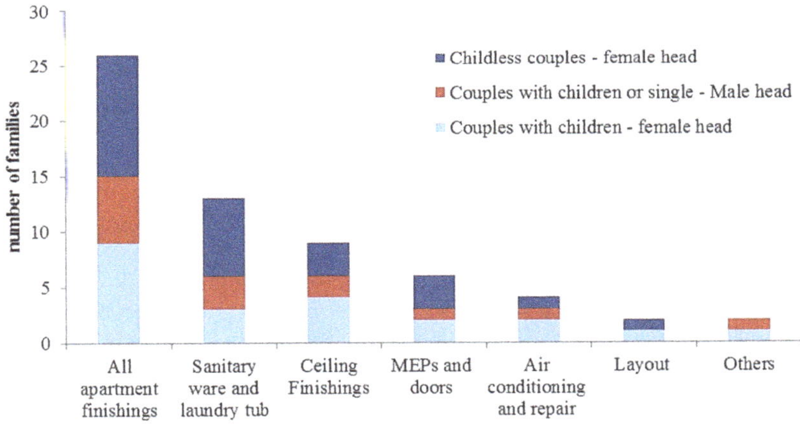

Figure 6. Correspondence analysis of the client profile clusters in relation to product changes made by users.

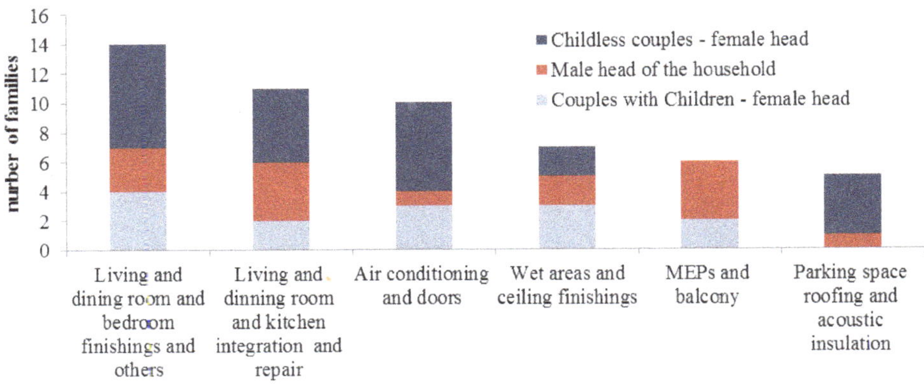

Figure 7. Correspondence analysis of client profile in relation to product changes still desired by users.

5.5. Hierarchical Value Maps

Figure 8 presents a HVM for the two house-building projects, considering the cut-off point of four relationships or 59% of the relationships. This map represents the main relationships between product attributes, consequences obtained in use, and expected values. For instance, it is possible to visualize the relationship between both floor and wall finishings in different rooms, in terms of the quality of materials, and how easy these are to clean. A major consequence is the hygiene and its impact on users' health, which increases the quality of living. The map also makes explicit the connection between users' complaints about the quality of the paint used, and the difficultly to clean the walls. Regarding the floor finishings, these were chosen based on how it feels, *i.e.*, tiled floors are connected with spacious sensation while wooden floors are connected with warmth. Those finishings mainly contribute to make a pleasant environment in terms of aesthetics, tranquility, and identity, leading to a better quality of life.

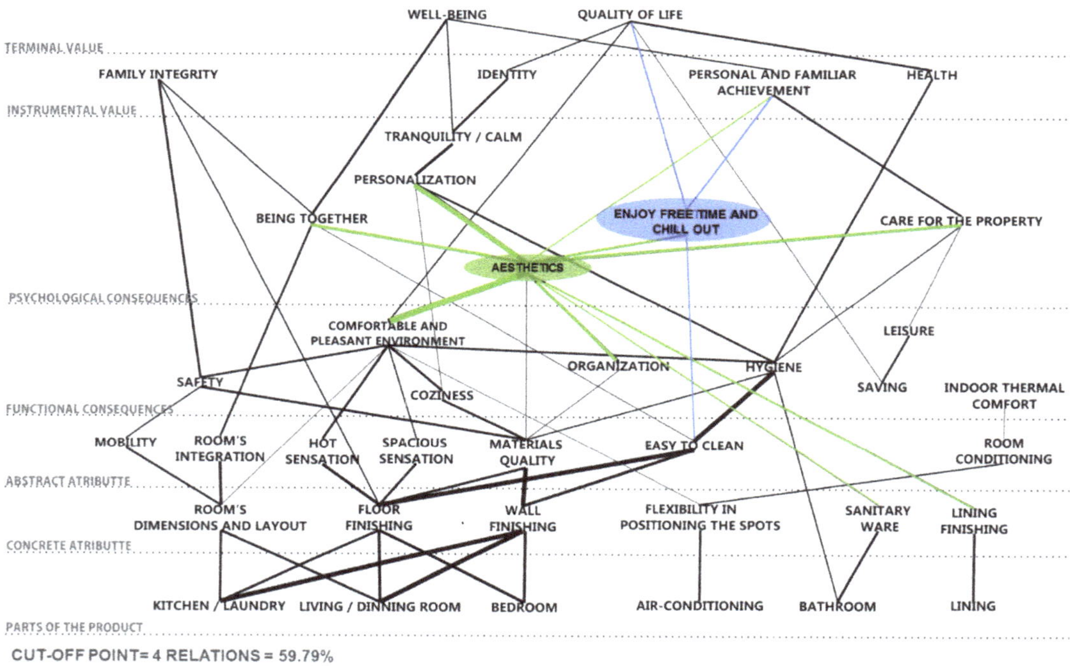

Figure 8. General hierarchical value map.

The items that generated dissatisfaction due to the lack of design flexibility also appear in the map. One of those items is the desire to integrate the living and dining room with the kitchen, shown in the map by the strong relationship between these two environments, and by the abstract attribute of dimension and the layout of the rooms, followed by integration. This customization is concerned with social interaction, resulting in the family well-being and integrity. Another source of dissatisfaction was the difficulty to install additional air conditioning units, which is related to thermal comfort and, subsequently, quality of living. Moreover, the problems related to the quality of execution and materials were mostly associated to the family's safety and physical integrity.

It is worth pointing out that the construct aesthetics is a central element in the map shown in Figure 8 (in green). From one hand, it is an important reason for customizing several attributes, such as wall and floor finishings, ceiling, and sanitary ware. On the other hand, it has a strong contribution to the achievement of different abstract goals, psychosocial consequences and values, such as personalization, care for the acquired property, personal and family fulfillment, and social interaction. The construct "enjoy free time and chill out" was repeatedly cited in connection to comments about a hectic daily life and working hard to be able to buy an apartment (connections in blue in Figure 8). Therefore, it is closely related to personal and family fulfillment, as well as with the quality of living. Two terminal values mentioned by interviewees, well-being and quality of living, seem to have similar meanings. However, in this context, the respondents made a distinction: quality of living was referred as an improved and lasting future condition, while well-being was related to feeling good momentary.

5.6. Summary Choice Menu

Table 7 presents a summary choice menu, which contains the list of the most relevant customization units and the class of items for each one of them, connecting them to the client profiles that should be targeted. In this set of customization units only those with a frequency above 10% are included, as suggested by Schoenwitz, Naim and Potter [38]. It also includes the customizations desired by over 20% of respondents.

Moreover, the summary choice menu also points out the main problems related to each customization unit from the point of view of the users, which can be interpreted as quality improvement opportunities. A regression analysis was performed with the aim of identifying the customization units that most contribute to overall satisfaction, in order to identify improvement opportunities. The satisfaction assessment of attributes were used as explicative variables and the overall satisfaction assessment as the response variable, resulting in an adjusted $R^2 = 0.7359$. The results indicated that there was a significant impact of the floor, wall and ceiling finishings (estimated coefficient = 0.49, $p = 0.000141$), and kitchen, living and dining room layout (estimated coefficient = 0.45, $p = 0.009$) on the overall degree of satisfaction. R^2 is the coefficient of determination, *i.e.*, a statistical measure of how well the regression line approximates the real data points, while p is the estimated probability of rejecting the null hypothesis when that hypothesis is true [37]. Therefore, each point on the rating assigned to those items in the individual satisfaction assessment is likely to raise the overall satisfaction level in nearly half a point.

The floor and wall finishing customization units were the most frequently ones carried out in all rooms of the housing units, despite the fact that changes in the specification of painting was not offered by the building company. It is important to highlight that the areas of social interaction where most daily activities take place were the ones with the highest number of customizations. Hence, the house-building company should focus on the quality of those elements and on offering choices with the aim of achieving a pleasant and hygienic environment, with the appearance desired by the customers, in order to achieve the goals of health, well-being and quality of life.

The customization units related to the ceiling finishings and the sanitary ware should be targeted to female heads of household, and their demands increases according to family income. As previously mentioned, this client profile is more concerned with home appearance. With regards to sanitary ware, the company should consider suppressing these components as an option for this customization unit, enabling the client to choose options that are not offered, leaving for the client to purchase and install it after project delivery. This strategy would avoid wasting materials and time for options that do not lead to customer satisfaction.

The customization units concerned with electricity services and air conditioning are primarily carried out by female heads of household. Regarding electricity services, the company offered the option of changing the voltage, because many buyers come from other towns that had a different voltage, so that the buyers do not need to spend money on replacing domestic appliances.

Table 7. Summary choice menu.

Customization Units	Options	Client Profiles	Importance	Strong Correlation to Overall Satisfaction	Improvement Opportunities
Floor Finishings	Ceramic floors	Couple with children—female as head of household	Hygiene → Health → Quality of life	x	Living and dining room and bedrooms are delivered without floor finishings;
		Couple with children or single—male as head of household	Spacious sensation→ Comfortable and pleasant environment → Aesthetics → Tranquility		
	Wood veneer floor	Couple with children and childless couple—female as head of household	Sensation of warmth → Comfortable and pleasant environment → Aesthetics → Being together → Well-being		
Wall Finishings	Painting	Couple with children—female as head of household	Hygiene → Health → Quality of life	x	Poor quality of the painting. It comes of when cleaning.
	Ceramic (wet areas)	Couple with children or single—male as head of household; Childless couple—female as head of household	Easy to clean → Comfortable and pleasant environment → Enjoy free time and chill out → Being together; Quality of the materials → Hygiene → Comfortable and pleasant environment → Aesthetics → Well-being		
Ceiling	Standard Gypsum liner	Female as head of household. The higher the income higher the frequency	Aesthetics → care for the property → Well-being	x	
Sanitary Ware	Sink (removal) Toilet	Couple with children and childless couple—female as head of household	Aesthetics → Well-being		Room too small, causing difficulties to place sanitary ware
Electrical installation	Change in voltage	Couple with children—female as head of household, and singles. The higher the income the higher the frequency	Saving → Comfortable and pleasant environment → Quality of life		Poor quality of the wiring and plugs.
	Flexibility in positioning spots		Comfortable and pleasant environment		
Air conditioning	Number of spots	Couple with children and childless couple—female as head of household	Room conditioning→ Comfortable and pleasant environment → Well-being		Impossibility to add new spots
Kitchen, living and dining room layout	Segregated			x	Kitchen is too small
	Kitchen/dining room	Couple with children or single—male as head of household; and Couple with children—female as head of household	Rooms integration → Being together →Family integrity → Well-being	x	

The only customization unit that depends on the layout of the housing unit is the integration of the kitchen with dining and living room. Although the building company offered that option at the design stage, many users reported that they did not know about it, and once the unit was delivered it was no longer possible to do it. No changes in that requirement have been made after delivery due to the need of changing structural elements of the building. In fact, Schoenwitz, Naim and Potter [38] state that customers tend not to change structural elements of their homes, due to risks associated with it, but they prefer to change those elements in which they can imprint their lifestyle and personal values, such as finishings. Hence, if the company explicitly offers that customization unit, users tend to be more confident that this will be done safely, and adding more value for them.

5.7. Evaluation of the Method by the Company

In the meeting carried out at the construction company for the evaluation of the method, three guiding questions were made to the participants:

(i) How useful was the information provided in terms of planning the space of solutions in the development of new projects?
(ii) How relevant and clear was the information about customization units?
(iii) How easy for the company was to use visual devices summarizing relevant information about the demands for customization?

The reaction of the participants in relation to the results of the implementation of the method was in general positive. They pointed out that the visual devices were useful not only for defining the customization units, but also for supporting other design decisions. For instance, the discussion was useful for the company in terms of deciding whether the laundry area should be an open area or not. The application of the method indicated that, although the company had plenty of data on customer requirements and satisfaction, these were not appropriately processed and analyzed.

A major interest of the company in the information produced was the complaints that came out in this investigation, and to check whether all the problems had been solved. This indicated that information on product changes carried out or desired by the user can be regarded as additional requirements or sources of dissatisfaction that have not been properly communicated through other channels. For a company that delivers customized products, those quality problems should be addressed before offering further choices, as suggested by Pine II [8], Barlow [5] and Svensson and Barfod [18].

The participants also pointed out that there was a misalignment between the purpose of the funding programs and what the clients expected in this type of house-building project. They suggested that the large number of product changes demanded or desired by customers might be caused by the fact that the decision to purchase a housing unit in those projects were mostly due to the favorable funding conditions, such as low interest rates, long time for paying the mortgage, rather than the housing product itself. For instance, many clients would like to have a balcony and a barbecue place, which is not typical for this type of funding program. This indicates that understanding the diversity of requirements and the demands for customization represent improvement opportunities not only for the product development process of house-building companies but also for the conception of governmental funding programs.

Finally, although the participants found useful the information on customization demands, they think that the company is not capable of delivering all customization units to the clients, due to limitations of the existing production control system, which is not capable of managing a very large space of solutions, in their opinion. They are very much concerned with the possibility of increasing costs beyond what is required by funding programs. Moreover, some of the customization units could not be offered due to the fact that the company needs to deliver all housing units complete at the end of the contract with the funding agency—for instance, a housing unit could not be delivered without all the sanitary ware installed, although some customers would prefer that.

6. Discussion and Conclusions

This paper has proposed a method for identification of the customization demands in low-cost housing. Its main practical contribution is a step-by-step process to identify relevant customization units and connect them with the benefits pursued by customers. By segmenting its market according to customer profiles, a housing-building company should be able to identify the most relevant product attributes to specific customers, and use that information as an input to analyze the feasibility of customization units, considering organizational capabilities. The outcomes of the method can also be used to structure the information that will be disseminated to potential customers. Therefore, a set of customization units can be chosen to be offered to potential customers that have the same profile, through the interpretation of bivariate analyses concerned with customization clusters and client profile clusters.

This investigation has made a contribution in terms of using MEC for a complex product such as social housing, which has not been investigated in previous studies. Such complexity demands an additional effort for clarifying the relationship between the parts of the product and the customization units, consequences in use and the desired values. In fact, one of the adaptations that was necessary in the MEC conceptual model, was the inclusion of an additional category at the bottom of the hierarchy, named product parts, which are connected to different product attributes. In order to deal with the complexity of considering so many attributes in the hierarchical value map, only the most relevant customization units were addressed in this application. In this respect, the laddering technique was useful for establishing a cut-off point, and choosing the most relevant links to be included in the map.

This method should be applicable to any type of housing market, although it seems to be much more relevant to the low-cost market segment, due to the obvious constraints in terms of costs. Further research is necessary to adapt this method to other types of building projects, such as commercial, industrial, and health care, since the data collection procedures and statistical analysis techniques were designed for the typical characteristics of low-cost housing projects, such as multiple housing units, subsidized funding, and wide range of customer profiles.

This investigation has also made a number of contributions related to the reasons why low-cost housing customers would like to change their units. For example, all clients have a declared preference for finishings that are easy to clean, so that the home hygiene is easy to be kept, contributing for the good health conditions, and the quality of living of the family, confirming the results of the study carried out by Zinas and Jusan [39]. The results obtained in the two projects have also pointed out the central role of aesthetics. Most users wanted to change the finishings of the units, including walls, floor, ceiling, and sanitary ware, in order to achieve some desired end states that can be explained by a set of very abstract constructs, such as identity, personal and familiar achievement, well-being and quality of life.

However, some customization demands are strongly related to specific clusters of customer profiles. For instance, the desire for a balcony in the apartment for leisure was made explicit only by male heads of the household. Moreover, this type of client tends to prefer a ceramic floor to create a spacious sensation, in order to have a comfortable and nice home. By contrast, for childless couple with a female as the head of the household, it is more important to integrate the living and dining room with the kitchen, and paint those rooms, in order to offer a pleasant personalized environment for socializing. Although the desire for personalization in housing in connection with the household lifestyle was already pointed out by Ozaki [40] and Schoenwitz, Naim and Potter [38], the application of the proposed method provide insights on which types of personalization are the most important ones and why.

A set of visual devices were proposed to disseminate the outcomes of the method, such as hierarchical value maps and the summary choice menu, which can be support decision-making in the product development process, or be used as templates for evaluation exercises.

The method was applied in two similar projects from the same company, with the aim of testing the method from the perspective of a single organization. In future studies, the method could be tested in a larger sample of projects, with the aim of carrying an analysis for segments of the low-cost housing market. Another possibility for future studies is to fully implement the method in the context of an organization in order to further evaluate its utility and applicability.

Acknowledgments

The authors would like to thank the National Council for Scientific and Technological Development (CNPq) and the Ministry of the Cities for the research grant that has partially funded this research work (Process No. 550274/2012-0).

Author Contributions

All four authors were involved in the design of the research work. Cynthia Hentschke carried out data collection and processing. Márcia Echeveste was strongly involved in the statistical analysis. Cynthia Hentschke, Carlos Formoso and Cecília Rocha wrote the paper. All four authors revised, read and approved the final manuscript.

Conflicts of Interest

The authors declare no conflict of interest.

References

1. Noguchi, M. The effect of the quality-oriented production approach on the delivery of prefabricated homes in Japan. *J. Hous. Built Environ.* **2003**, *18*, 353–364.
2. Brandão, D.Q.; Heineck, L.F.M. Significado multidimensional e dinâmico do morar: Compreendendo as modificações na fase de uso e propondo flexibilidade nas habitações sociais. *Ambiente Construído* **2003**, *3*, 35–48.
3. Formoso, C.T.; Leite, F.; Miron, L.I.G. Clients requirement management in social housing: A case study on the residential leasing program in Brazil. *J. Constr. Dev. Ctries* **2011**, *16*, 47–67.
4. Noguchi, M.; Hernández-Velasco, C.R. A mass custom design approach to upgrading conventional housing development in Mexico. *Habitat Int.* **2005**, *29*, 325–336.
5. Barlow, J. From craft production to mass customisation innovation requirements for the U.K. housebuilding industry. *Hous. Stud.* **1999**, *14*, 23–42.
6. Hart, C.W.L. Mass customization: conceptual underpinnings, opportunities and limits. *Int. J. Serv. Ind. Manag.* **1995**, *6*, 36–45.
7. Jiao, J.; Ma, Q.; Tseng, M.M. Towards high value-added products and services: Mass customization and beyond. *Technovation* **2003**, *23*, 809–821.
8. Pine, B.J., II. Mass Customizing Products and Services. *Strategy Leadersh.* **1993**, *21*, 6–55.
9. Da Silveira, G.; Borenstein, D.; Fogliatto, F.S. Mass customization: Literature review and research directions. *Int. J. Prod. Econ.* **2001**, *72*, 1–13.
10. Barlow, J.; Ozaki, R. Achieving customer focus in private housebuilding: Current practice and lessons from other industries. *Hous. Stud.* **2003**, *18*, 87–101.
11. Van den Thillart, C.; Oostra, M. Mass Customisation a new Challenge for the Building Sector—Concepts and Networks for the European Market. In *New Perspective in Industrialisation in Construction*; A State-of-the-Art Report; Girmscheid, G., Scheublin, F., Eds.; CIB Publication 329 (CIB TG57): Zurich, Switzerland, 2010.
12. Rocha, C.G. A Conceptual Framework for Defining Customisation Strategies in the House-Building Sector. Ph.D. Thesis, Federal University of Rio Grande do Sul, Porto Alegre, Brasil, 2011.
13. Tillmann, P.A.; Formoso, C.T. Opportunities to adopt mass customization—A case study in the Brazilian house building sector. In Proceedings of the IGLC 16, Manchester, UK, 16–18 July 2008.
14. Linner, T.; Bock, T. Evolution of large-scale industrialization and service innovation in Japanese prefabrication industry. *Constr. Innov.* **2012**, *12*, 156–178.
15. Yashiro, T. Conceptual framework of the evolution and transformation of the idea of the industrialization of building in Japan. *Constr. Manag. Econ.* **2014**, *32*, 16–39.
16. Smith, S.G. The essential qualities of a home. *J. Environ. Psychol.* **1994**, *14*, 31–46.
17. Williamson, P.; Westbrook, R. Mass Customization: Japan's New Frontier. *Eur. Manag. J.* **1993**, *11*, 38–45.
18. Svensson, C.; Barfod, A. Limits and opportunities in mass customization for "build to order" SMEs. *Comput. Ind.* **2002**, *49*, 77–89.

19. Fogliatto, F.S.; da Silveira, G.J.C. Mass customization: A method for market segmentation and choice menu design. *Int. J. Prod. Econ.* **2008**, *111*, 606–622.
20. Fischer, S.; Santos, A. dos. Diretrizes de Projeto Arquitetônico Para Permitir a Expansão de Habitações de Interesse Social. In Proceedings of the III Simpósio Brasileiro de Gestão e Economia da Construção, UFSCAR, São Carlos, Brazil, 16–29 September 2003. Available online: http://www.deciv.ufscar.br/sibragec/apresentacao/index.htm (accessed on 1 January 2013).
21. Gutman, J. A Means-End Chain Model Based on Consumer Categorization Processes. *J. Mark.* **1982**, *46*, 60–72.
22. Gengler, C.E.; Klenosky, D.B.; Mulvey, M.S. Improving the graphic representation of means-end results. *Int. J. Res. Mark.* **1995**, *12*, 245–256.
23. Zeithaml, V.A. Consumer perceptions of price, quality, and value: A means-end model and synthesis of evidence. *J. Mark.* **1988**, *52*, 2–22.
24. Woodruff, R.B.; Gardial, S. *Know Your Customer: New Approaches to Understanding Customer Value and Satisfaction*; Wiley: Malden, MA, USA, 1996; p. 360.
25. Leppard, P.; Russell, C.; Cox, D. Improving means-end-chain studies by using a ranking method to construct hierarchical value maps. *Food Qual. Pref.* **2004**, *15*, 489–497.
26. Reynolds, T.J.; Gutman, J. Laddering theory, method, analysis, and interpretation. *J. Advert. Res.* **1988**, *Feb/March*, 11–31. Available online: http://www.uta.edu/faculty/richarme/MARK%205338/Reynolds%20and%20Gutman%20laddering%20article.pdf (accessed on 1 January 2013).
27. Russell, C.G.; Flight, I.; Leppard, P.; van Lawick van Pabst, J.A.; Syrette, J.A.; Cox, D.N. A comparison of paper-and-pencil and computerised methods of hard laddering. *Food Qual. Prefer.* **2004**, *15*, 279–291.
28. Bonatto, F.S.; Miron, L.I.G.; Formoso, C.T. Evaluation of social housing projects based on user perceived value hierarchy. In Proceedings of 19th Annual Conference of the International Group for Lean Construction (IGLC), Lima, Peru, 13–15 July 2011.
29. Yang, B.; Burns, N. Implications of postponement for the supply chain. *Int. J. Prod. Res.* **2003**, *41*, 2075–2090.
30. Blecker, T.; Abdelkaf, N. Complexity and variety in mass customization systems: Analysis and recommendations. *Manag. Decis.* **2006**, *44*, 908–929.
31. Reynolds, T.J.; Olson, J.C. Understanding consumer decision making: The means-end approach to marketing and advertising strategy. In *Applied Health Economics and Health Policy*; Lawrence Erlbaum Associates: London, UK, 2008.
32. Overby, J.W.; Gardial, S.; Woodruff, R.B. French *versus* American Consumers' Attachment of Value to a Product in a Common Consumption Context: A Cross-National Comparison. *J. Acad. Mark. Sci.* **2004**, *32*, 437–460.
33. Barrena, R.; Sánchez, M. Consumption frequency and degree of abstraction: A study using the laddering technique on beef consumers. *Food Qual. Prefer.* **2009**, *20*, 144–155.
34. Lukka, K. The constructive research approach. In *Case Study Research in Logistics*; Ojala, L., Hilmola, O.-P., Eds.; Publications of the Turku School of Economics and Business Administration: Turku, Finland, 2003; pp. 83–101.

35. March, S.T.; Smith, G.F. Design and natural science research on information technology. *Decis. Support Syst.* **1995**, *15*, 251–266.
36. Villa, S.B. Avaliando a habitação: Relações entre qualidade, projeto e avaliação pós-ocupação em apartamentos. *Ambiente Construído* **2009**, *9*, 119–138.
37. Hair, J.F., Jr.; Black, W.C.; Babin, B.J.; Anderson, R.E. *Multivariate Data Analysis*, 7th ed.; Pearson Prentice Hall: Upper Saddle River, NJ, USA, 2009.
38. Schoenwitz, M.; Naim, M.; Potter, A. The nature of choice in mass customized house building The nature of choice in mass customized house building. *Constr. Manag. Econ.* **2012**, *30*, 37–41.
39. Zinas, B.Z.; Jusan, M.B.M. Housing choice and preference: Theory and measurement. *Procedia* **2012**, *49*, 282–292.
40. Ozaki, R. “Mind the gap”: Customers’ perceptions and the gaps between what people expect and what they are offered. In *Consumer Choice in Housing—The Beginnings of a House Buyer Revolt*; Joseph Rowntree Foundation: York, UK, 2002; pp. 45–67.

Retrofitting Housing with Lightweight Green Roof Technology in Sydney, Australia, and Rio de Janeiro, Brazil

Sara Wilkinson and Renato Castiglia Feitosa

Abstract: The built environment contributes around half of total greenhouse gas emissions and with 87% of residential buildings that we will have by 2050 already built, it is vital to adopt sustainable retrofitting practices. The question is: what are the viable solutions? One answer may be green roof retrofitting. The environmental benefits include reduced operational carbon emissions, reduced urban heat island effect, increased bio-diversity, housing temperature attenuation and reduced stormwater run-off. The economic benefits are the reduced maintenance costs and lower running costs. The social gain is the creation of spaces where people have access to green areas. However, the barriers to retrofitting include the perceptions of structural adequacy, the risk of water damage, high installation and maintenance costs, as well as access and security issues. Many Australian and Brazilian residential buildings have metal sheet roofs, a lightweight material with poor thermal performance. During the summer, temperatures in Sydney and Rio de Janeiro reach 45 degrees Celsius, and in both cities, rainfall patterns are changing, with more intense downpours. Furthermore, many residential buildings are leased, and currently, tenants are restricted by the modifications that they can perform to reduce running costs and carbon emissions. This research reports on an experiment on two small-scale metal roofs in Sydney and Rio de Janeiro to assess the thermal performance of portable small-scale modules. The findings are that considerable variation in temperature was found in both countries, indicating that green roof retrofitting could lower the cooling energy demand considerably.

Reprinted from *Sustainability*. Cite as: Wilkinson, S.; Feitosa, R.C. Retrofitting Housing with Lightweight Green Roof Technology in Sydney, Australia, and Rio de Janeiro, Brazil. *Sustainability* **2015**, *7*, 1081-1098.

1. Introduction

There is a consensus among climate scientists that global weather patterns are changing, with some regions getting hotter and drier, while others will become wetter [1,2]. One of the highest contributors to human-induced climate change is the built environment [2]. Within the built environment, the biggest land use type contributing to greenhouse gas emissions is the residential sector [3]. While efforts are being made globally to improve sustainability in buildings through operational and embodied energy-efficient design, most of the stock that will exist by 2050 is already here. It is estimated that 87% of residential buildings the U.K. will have in 2050 been built already [4]. With many existing residential buildings leased and home ownership rates in decline in many countries, as well as many other residential buildings that could benefit from a low-cost, easily retrofitted means of reducing energy costs and carbon emissions, this exploratory research is a worthwhile undertaking.

In addition, many cities are experiencing rapid urban expansion and/or densification, which contributes to the urban heat island effect, whereby city centres are sometimes up to five degrees warmer than the outer suburbs [5]. In addition, within some cities, urban heat canyons are created, so that heat is trapped between buildings, which can contribute to negative human health impacts and even fatalities during excessively hot days [6,7]. In the recent past, high temperatures have been observed worldwide; Rio de Janeiro experienced historical record high temperatures in February, 2014 [8]. In a four- to five-day period of plus 45 degree Celsius days in January, 2014, in Melbourne, Australia, more than twice the average rate of mortality was experienced. These deaths were attributed to the excessive heat conditions, which were exacerbated in the city centre, where heat was trapped between buildings and under tree cover in an urban canyon [9]. With given predicted climate change impacts and an ageing population, these figures look likely to increase. On this basis, the focus for climate change mitigation is through adaptation and sustainable retrofitting of existing buildings. In addition, the statistics of increasing temperatures and ageing populations are broadly similar across many countries, and retrofitting buildings has universal application. The question is: what are the viable solutions in terms of retrofitting our existing residential buildings? Here, viable is defined as the retrofit measures that are environmentally, economically and socially worthwhile to undertake.

If the aim is to reduce building-related green-house gas emissions, for some regions, the goal will be to keep buildings cool and, therefore, to reduce cooling loads, whilst for other areas, the problem will be one of retaining heat and reducing heat loss through leaky buildings. Whereas in other regions, the problem will be one of accommodating increased frequencies of intense rainfall [5]. One answer, which may suit a number of regions, is to retrofit buildings with green roofs, as there are environmental, economic and social benefits.

The environmental benefits include potential reductions in operational carbon emissions, reductions in the urban heat island, increases in bio-diversity, housing temperature attenuation and reductions in stormwater run-off [10–12]. Air quality is improved as plants remove carbon dioxide and harmful pollutants from the atmosphere, and in addition, green roofs provide habitats for insects, birds and reptiles to shelter and find food and water [13].

Thermally, the mass of the green roof improves the insulating qualities of the building by reducing heat transmission through the roof. Much heat loss occurs through the roof as heat rises and then escapes through inadequately-insulated and poorly-sealed roof structures. Some authors have evaluated the role of green roofs' cooling and warming potential in energy savings and the potential for retrofitting, based either on modelling or experimental data [10]. There is consensus that, in non-insulated buildings, which are common in Rio de Janeiro and Sydney, green roofs can improve the insulation properties and reduce annual energy consumption. According to Castleton [10], over the past 10 years, several studies have shown that green roofs can offer benefits in winter heating reduction, as well as summer cooling. Nichaou *et al.* [14] showed an annual energy saving potential of green roofs on non-insulated buildings for heating of 45%–46% and for cooling of 22%–45%. Wong *et al.* [15] found an annual energy savings of 10.5% for an un-insulated extensive green roof covered in turf, compared with a non-greened un-insulated roof. However, the energy savings are realised only on floors near the roof. Alcazar and Bass [16] state that due to the tall nature of the buildings, roofs comprise around 16% of the total building envelope, and the largest reductions in energy consumption were seen in rooms directly below the green roof. No energy savings were found more than three floors below the roof [16].

Where stormwater or pluvial flooding is an issue, green roofs can reduce the run-off rate and also filter or cleanse the water passing through the roof covering [5]. There are numerous environmental benefits from the installation of green roofs in urban settlements, which are suitable whether the problem is one of excess stormwater or the need to enhance thermal performance. It is the case that the specification of white roofs, roofs that are painted white or with reflective colours, is the most cost-effective means of reducing the heating load in buildings; however, bio-diversity and air quality benefits are absent with this option [17]. The decision to retrofit a green roof has multiple variables and should not be evaluated on one variable alone, but on the multiple benefits that are delivered [18].

Economically, the benefits to occupiers and owners are reduced roof maintenance costs and lower running costs [10]. There are erroneous perceptions, however, among the practitioner community that green roofs lead to higher maintenance costs [18], which is resulting in less application of green roof technology in buildings.

The third aspect, the social gain, is the creation of spaces where people have greater access to nature. The biophilia effect describes the phenomenon in which humans experience positive feelings as a result of the connection to the natural environment [19]. Unfortunately, for many city residents, access to the natural environment is limited and diminishing [12]. In Sydney, for example, it is estimated that there are less than 22 square metres per resident and that only about 15.5% of the city is covered by urban canopy [12]. The city wishes to increase this level of urban greenery for the health and well being of the community [12]. There are initiatives seeking to increase the amount of urban greenery in Sydney by twenty percent before 2020 [20], and the application of green roofs would be a way of contributing to this target. On the contrary, the amount of green spaces in Rio de Janeiro has decreased significantly, mostly due to the lack of space and population growth [21]. No plans have been adopted yet to deal with this problem.

However, there are barriers to the adoption of retrofitted green roofs, which include perceptions of structural adequacy, risk of water damage, high installation and maintenance costs, as well as access and security issues [11].

In some locations, such as Sydney, the intent will be to reduce cooling loads, whereas others locations, such as London, will desire thermal insulation or reduction in stormwater run-off. The ability to meet the demands will depend on the available budget and physical characteristics. Although the technology to design and retrofit green roofs exists, the uptake and the demand have not been high. Overall, the gains have not been deemed sufficient, and in both cities, the existing numbers of residential green roofs confirm this observation. In 2014, the City of Sydney adopted the first green roofs and walls policy for Australia, which sets out a commitment to increase the number of high quality green roofs and walls [22]. The policy includes a three-year implementation plan to ensure that the policy is understood, properly adopted and integrated. There are 59 green roofs in Sydney currently, which serve a variety of purposes, including enhancing thermal performance [12].

Many Australian and Brazilian residential buildings have profiled metal sheet roofing; a lightweight material with poor thermal performance, where the heat transfer is very high. Many buildings have little or no insulation to offset the high heat gains. During summer periods, Sydney and Rio de Janeiro, experience temperatures that can reach 45 degrees Celsius and rainfall patterns that are variable and changing, affected by La Niña and Il Niño weather cycles. This research reports on an experiment on two small-scale profiled metal sheet roofs in both cities to assess thermal performance. Metal roof coverings are fairly typical in Australian housing design and have less load bearing capacity than other coverings, and a lightweight, low-cost, easily-installed module might be a good option for retrofitting; the question that needs addressing is: how do these modules perform? In each city, one roof was left as a control, whilst the second roof was planted with succulent plants in lightweight trays. Data were collected using thermal data loggers over a summer and autumn season. The paper discusses the findings and the potential for retrofitting residential stock with lightweight trays planted with succulents.

2. Methodology

The methodology adopted is predicated on the development of simple technologies to mitigate the problems created by increasing urban densification, which exacerbates urban heat islands and contributes to uncomfortably high internal housing temperatures. There are many technologies and approaches available to execute this research, but in this case, the researchers aimed to use adaptive techniques that minimised initial costs and maintenance costs; in other words technologies that would be affordable and easy to implement. For this reason, this project used lightweight removable modules of vegetation (rectangular containers) of a low thickness. This modular system enables planting, cultivation and maintenance off-site to be undertaken. This is an exploratory study to assess the thermal performance and feasibility of retrofitting with low-cost, lightweight roofing modules for metal sheet roofing.

The researchers sought to evaluate the performance of a green roof retrofit system, which could be widely used in metropolitan areas. At this point, there is an absence of empirical evidence on the performance of green roofs in Australia, with most data coming from the U.S. or Europe, where climatic conditions are very different [23]. Similar conditions exist for South America in terms of empirical data on green roof performance. Previous studies have shown significant variations based on temperatures, evaporation rates and wind conditions, which affect the performance of green roofs, but they have used longer life and more permanent technologies [23].

Two experiments were performed; one in Australia (Sydney) and one in Brazil (Rio de Janeiro). The Australian site is located on the roof of a building at the University of Technology, Sydney, in Ultimo, and the Brazilian location is on the roof of an existing building at the Oswaldo Cruz Foundation (Fiocruz).

Rio de Janeiro (S 23°; W 43°) has a tropical wet and dry or savannah climate with dry winters (Aw), and Sydney (S 33.9°; W 151.2°) has a humid subtropical climate (Cfa). The high and low annual average temperatures in Rio de Janeiro are 27.3 and 21 °C, respectively, whereas for Sydney, these temperatures are 21.7 and 13.9 °C. However, extreme higher temperatures are more commonly observed in Sydney (39.8–48.8 °C), rather than in Rio de Janeiro (37.7–39 °C) [2].

Succulent plants, such as variegated sedum, *Echeveria glauca* and *Kalanchoe* quicksilver, were selected on the basis of their higher drought resistance qualities and a lower risk of fire. Furthermore, these species can develop easily in shallow soils, and therefore, importantly, structural reinforcement of existing roofs is unnecessary. Additionally, due to the modular characteristics of the planting containers, the modules can be applied directly onto the roof covering, be it profiled metal sheeting or tiles.

Rectangular plastic containers were selected according to the availability at the different sites (Rio de Janeiro, 400 mm × 500 mm, and Sydney, 190 mm × 330 mm) where the experiments were performed, as shown in Figure 1.

(a) (b)

Figure 1. Rectangular plastic containers used in the temperature experiments. (**a**) Brazilian module, provided by Cidade Jardim Institute; (**b**) Australian module.

Both containers have a water storage system, which meets two main objectives. Firstly, it provides water to the soil through evaporation, enhancing the plants survival, even during extended periods of no rainfall. Secondly, it can attenuate temperature fluctuation due to the water layer between the soil and the roof.

The soil is separated from the drainage system by a permeable fabric (Geotextile), which allows the passage of water, but prevents the soil from leaking into the water chamber. For the plant species used in this research, a soil with good drainage and low organic content was used. A composition of two parts of sand to one part of loam was employed.

The evaluation of the green roofs' cooling potential is performed by the comparison between two housing prototypes with vegetated and non-vegetated roofs. Due to financial limitations and because of the exploratory nature of the research, it was not possible to use full-scale housing for the experiment. Therefore, small-scale structures are used to demonstrate the thermal performance of a non-green, traditional roof and a green roof. The experimental set-up comprises covering the roof of one of the prototypes with planted soil containers. Different types of housing prototypes are considered in this study. As shown in Figure 2, the Rio de Janeiro tests were carried out using small blockwork houses covered with metal sheeting, whereas in Sydney, metallic sheds were employed. In Australian housing, profiled metal sheet roofs are typically specified, and for this reason, the metal sheds were selected.

A simultaneous comparison between the records of temperature inside the vegetated and non-vegetated structures was made using data loggers that collect continuous temperature records over long periods of time. The temperature measurements were carried out using Extech TH10 Temperature USB Data loggers, using a time sampling of 30 min. The Rio de Janeiro tests were performed over 194 days, from 17 October 2012, to 29 April 2013, whereas the Sydney trial tests comprised a 97-day period from 11 December 2013, to 18 March 2014.

The data loggers were positioned at different heights inside each of the prototypes, according to the experimental site. In Rio de Janeiro and Sydney, they were placed 250 and 50 mm, respectively, below the top of the structure. All of the temperature differences observed are attributed only to the influence of heat incidence on the structures, given that the weather conditions between the vegetated and un-vegetated prototypes are identical.

(a) (b)

Figure 2. Housing thermal experiments. (**a**) Rio de Janeiro; (**b**) Sydney.

3. Data Analysis and Discussion

The results with regards to the green roof cooling potential for the two experimental sites (Rio de Janeiro and Sydney) are shown in Figures 3 to 5 below. Notwithstanding some basic differences in the structures (that is, blockwork in Rio de Janeiro and metal sheeting in Sydney), the tendency in temperature attenuation is evident. The measurements performed in Rio de Janeiro and in Sydney show that green roofs are able to attenuate daily variations of temperature.

3.1. Rio de Janeiro Data

Figure 3 presents a comparison between the non-green and green roofs' internal temperatures, during the 194-day data collection period, which comprises the whole Brazilian summer period and also part of spring and autumn. Some of this work was partially reported in [24].

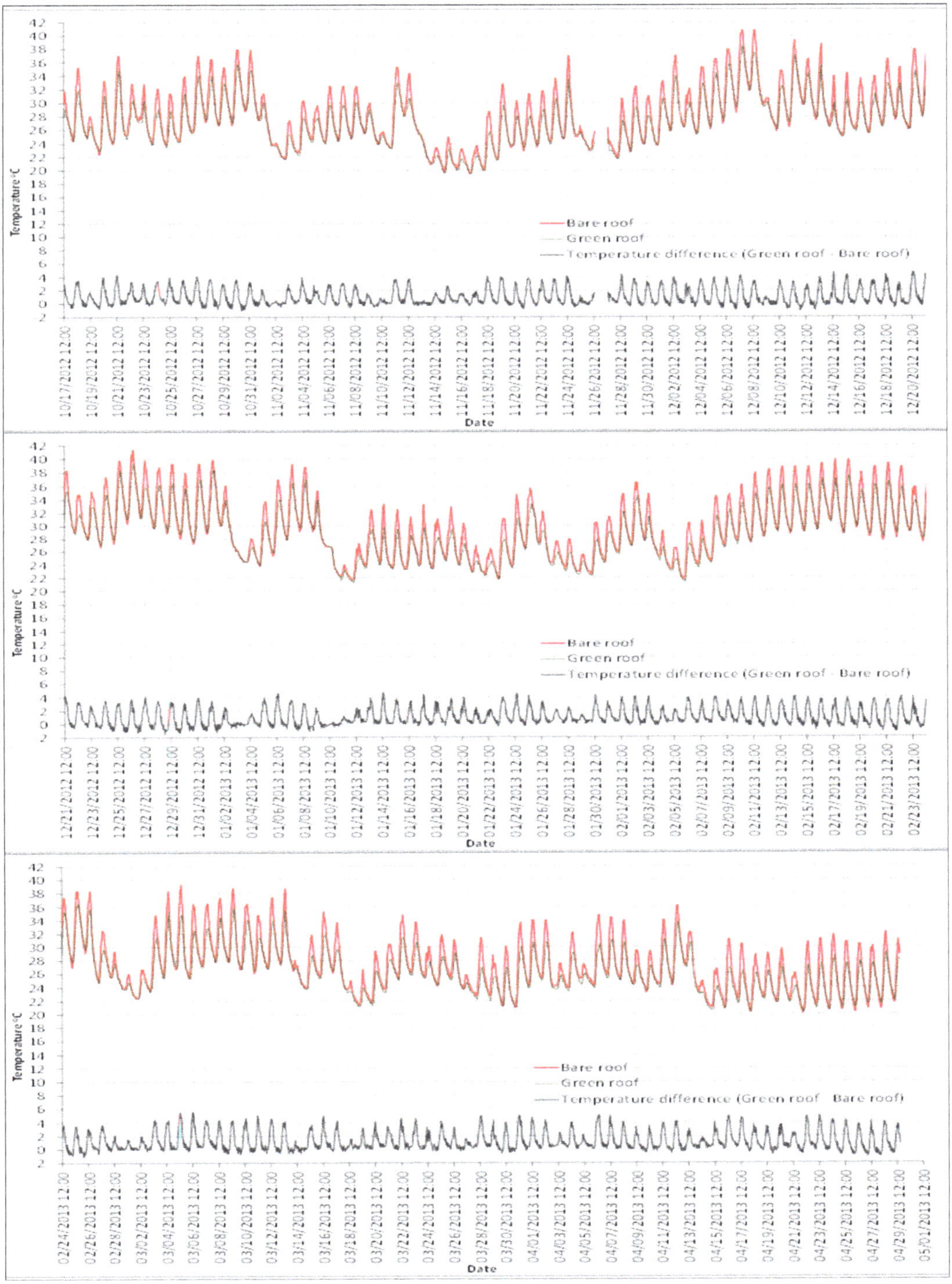

Figure 3. Comparison between the inner temperature of the non-green and green roofs, Rio de Janeiro, Brazil.

Figure 4 presents the typical detail of three out of the 194-day temperature comparison, where it mostly can be observed that the highest temperature differences are between 12:00 and 15:00.

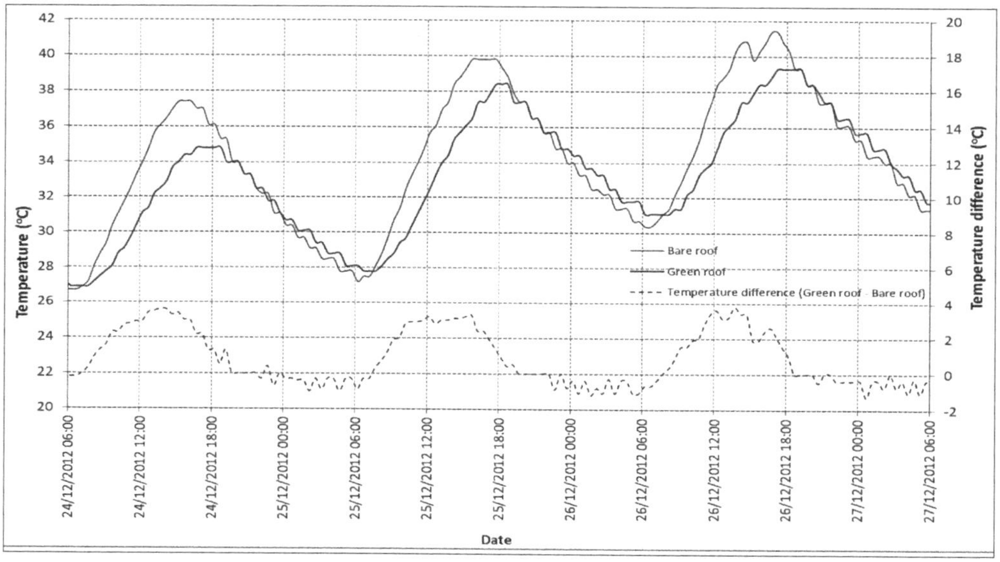

Figure 4. Typical detail of the temperature records between green and non-green (bare) roofs. Rio de Janeiro site.

During the whole period of investigation (194 days), the non-green roof presented maximum, minimum and average temperatures equal to 41.1 °C, 20.1 °C and 28.8 °C, respectively. Correspondingly, the values observed in the green roof case were 39.3 °C, 20.3 °C and 27.7 °C.

Based on daily variations of temperature, the maximum values observed during the daytime for non-green and green roofs varied from 23.9 °C to 41.4 °C and 23.2 to 39.3 °C, respectively. The minimum values that occurred during the nighttime varied from 20.1 °C to 31.8 °C for the non-green roof and 20.3 °C to 31.8 °C for the green roof. It is important to highlight that the green roof cooling potential, in attenuating high temperatures, is not directly related to the differences observed between those limits presented, due to the existing time lag between the non-green and green roof temperature peaks.

Comparing the simultaneous temperature differences between green and non-green roofs (Figure 1), it was observed that these values vary from −1.5 °C to 5.6 °C. The temperature differences were dependent of the temperature background. That is to say, that past temperature influences the present ones, showing that green roofs tend to buffer thermal exchanges. If, for example, a warm day increases the inner temperature, this internal heat will influence the following temperature. Positive values mean higher non-green roof temperature. The lowest positive temperature differences between non-green and green roofs (≤2 °C) were observed when the internal non-green roof temperatures were below 30 °C. The highest positive temperature differences (≥5 °C) were registered at the end of the summer period, when during the previous nighttime period, the green roofs' temperatures were cooler than the non-green roofs. However, higher green roof nocturnal temperatures contribute to weakening the following daytime temperature differences (<5 °C). The delay observed between the temperature peaks of non-green and green roofs results in slightly warmer green roof temperatures (negative differences) during the nighttime and early morning periods, which contributes to diminishing the temperature differences the following day.

3.2. Sydney Data

Despite the shorter period, when compared to Rio de Janeiro data, it was observed that a significant green roof cooling occurs. However, according to the characteristics of the site where the experiments were undertaken, a particular pattern in temperature registers can be observed. As shown in Figure 5, a sudden reduction in temperature occurs both for non-green and green roofs around 3 pm, due to the shading created by adjacent buildings. This figure shows an attenuation in temperature variation for green roofs, which is consistent with their insulating properties.

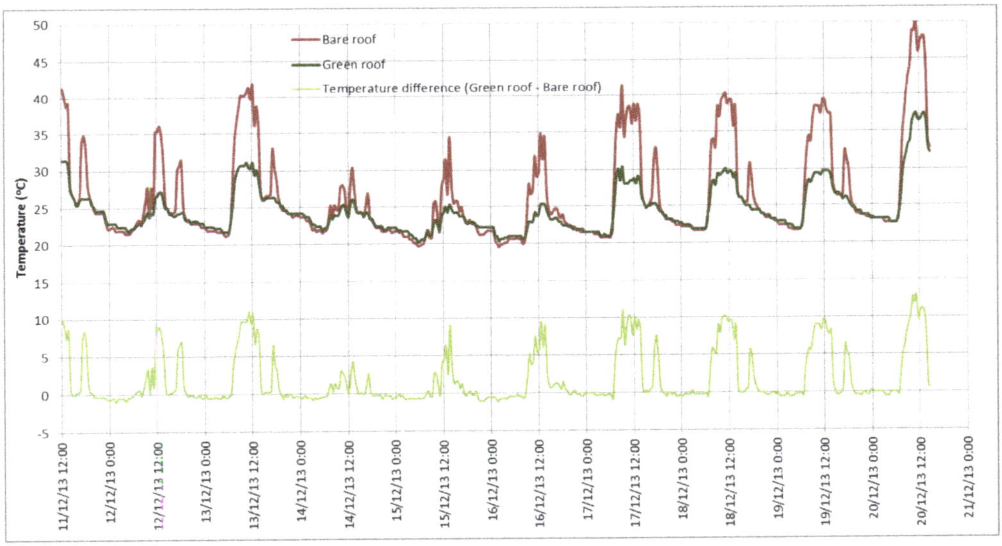

Figure 5. Influence of shadows on temperature caused by adjacent buildings for Sydney experimental roofs.

Figure 6 presents a comparison between non-green and green roof inner temperatures, for 133 days, during the summer period. The Sydney non-green roofs presented maximum, minimal and average temperatures equal to 50.3 °C, 17.2 °C and 25.2 °C, respectively. The values observed in the green roof case were 37.4 °C, 17.6 °C and 23.9 °C.

The temperature differences between green and non-green roofs varied from −1.6 °C to 14.8 °C. The lowest positive temperature differences between the non-green and green roofs (≤4 °C) were observed for the non-green roofs' inner temperature being under 30 °C. The highest positive differences (≥10 °C) occurred at non-green roof temperature peaks higher than 42 °C.

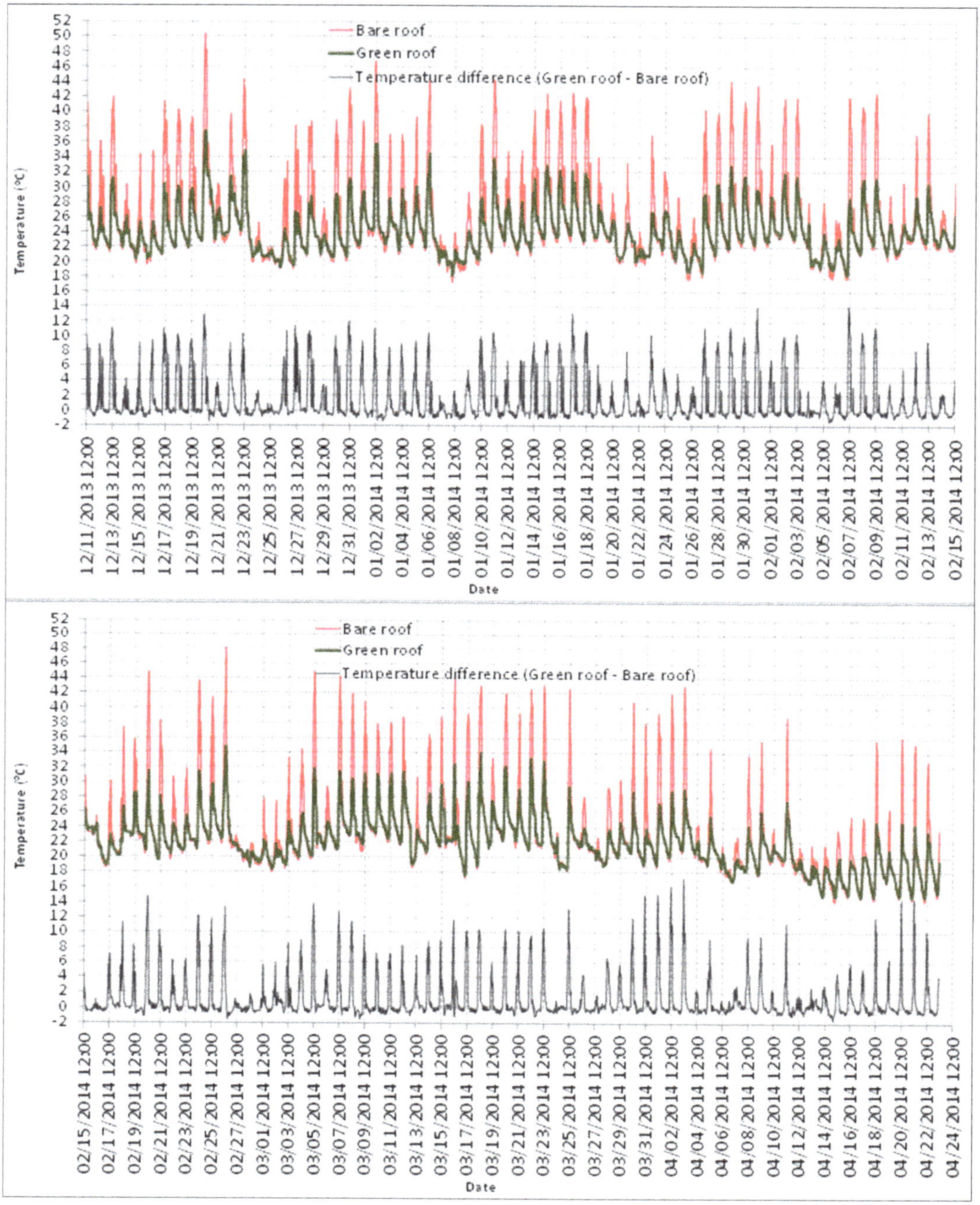

Figure 6. Non-green and green roofs' inner temperature comparison, Sydney, Australia.

Figure 7 depicts a two-day detail of the whole set of data presented in Figure 6. Similarly to what was observed at the Rio de Janeiro site, the highest differences in the temperatures between green and non-green roofs also were detected around noon. Negative differences were evident practically along all nighttime periods, additionally corroborating the efficiency of the green roof in attenuating high and relatively low temperatures.

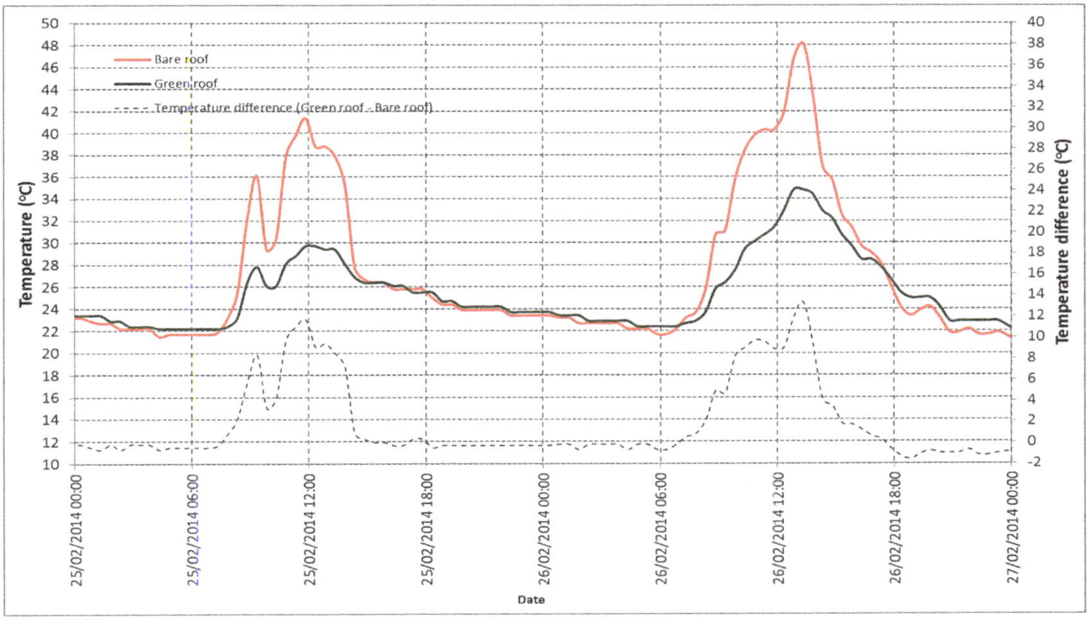

Figure 7. Typical detail of the temperature records between green and non-green roofs, Sydney site.

3.3. Evaluation of Rio de Janeiro and Sydney Data

This work does not solely intend to perform a comparison between the Sydney and Rio de Janeiro experiments, but it aims to evaluate the potential of the green roofs to attenuate housing temperature.

In both the Rio de Janeiro and Sydney cases, it could be seen that the green roof's capabilities lie in improving the insulation properties of the roof, due to the reduction of heat gain in the housing prototypes and, consequently, the temperature reduction, when compared to the non-green roof. In other words, the vegetation increases the thermal performance of the roof, which is quantified by the *U*-value parameter. This parameter quantifies the heat gain or loss through a surface and is also known as the overall heat transfer coefficient. A decrease in this parameter means that there is an improvement in thermal performance due to the attenuation of heat transfer into the buildings. According to work presented by Wong *et al.* [15], in a non-thermal insulating building, a green roof coverage with turf reduces the *U*-value from 2.39 to 1.19 W/m^{2}K.

Table 1 depicts the non-green and green roofs' maximum, minimum and average temperatures, as well as their higher and lower differences for the Sydney and Rio de Janeiro sites. It was observed that the green roof cooling potential in Sydney was greater than that in Rio de Janeiro, which was most likely due to the existing differences in both experiments. This may be attributed basically to the positioning of the temperature data loggers. In Sydney, they were located about 50 mm below the roof, whereas in Rio de Janeiro, they were 250 mm below the roof. As mentioned previously, the goal of this work is not the comparison between the two sites. Surrounding buildings for parts of the day overshadowed the Sydney site, and the amount and duration of overshadowing changes over time as the Sun's height increases or decreases. The study sought to evaluate the insulating properties of green roofs in different scenarios and under different criteria. This can be corroborated by temperature differences (up to 14.8 °C) observed in the Sydney experiments, taken so close to the inner roof surface, composed of a good thermal conductor (metal sheet).

Table 1. Experimental temperature comparison between Sydney and Rio de Janeiro.

Temperatures (°C)	Rio de Janeiro			Sydney		
	Non-green roof	Green roof	Simultaneous temperature difference	Non-green roof	Green roof	Simultaneous temperature difference
Maximum	41.4	39.3	5.6	50.3	37.4	14.8
Minimum	20.1	20.3	−1.5	17.2	17.6	−1.6
Average	28.8	27.7	-	25.2	23.9	-

According to Castleton [10], in addition to insulation, green roofs also add thermal mass, which is the material capacity to absorb and store energy, and thus, they provide inertia against temperature fluctuation, resulting in a temperature peak delay. However, analogous to Castleton [10], the green roof internal temperature peak delay observed is not relevant when compared to the non-green roof. In the present work, according to Figures 4 and 7, a slight time delay between green and non-green roof temperatures in both Sydney and Rio de Janeiro sites can be seen. The higher observed delays mostly ranged from 30 to 90 min, showing a more insulating, rather than a thermal mass insulating, property of the green roofs.

Besides the insulation properties of the green roofs, it is believed that their relevance in temperature attenuation and, consequently, energy savings in urban environments may be affected by neighbourhood shading conditions and different roof side conditions.

Another aspect to consider is related to the water and/or its levels in the soil and in the storage systems. Due to the high specific heat, water is supposed to provide inertia against temperature fluctuations. Although water levels have not been monitored in the current study, previous works have evaluated this effect [10,15,16,25,26].

Regarding the soil moisture content, studies performed by Alcazar and Bass [16] revealed that the thermal performance decreases for wetter soils, since water is a better conductor than air. However, Del Barrio [25] indicates the opposite, that is water promotes insulation. The results presented by Wong *et al.* [15] concur with Nichaou *et al.*, Alcazar and Bass [16] and Castleton *et al.* [10]: higher U-values (heat gain) being observed for higher water soil content.

Lazzarin *et al.* [26] presented an evaluation of passive cooling and the role of evapotranspiration, considering the wet and dry soil conditions of the green roofs. According to these authors, compared to a traditional roof, even under dry conditions, green roofs reduce the incoming heat flux by 60%. However, in a wet green roof, the additional evapotranspiration not only avoids heat entry, but also removes it from the building. The water effect in the green roofs' cooling potential seems to be dependent simultaneously on the soil water content and the evapotranspiration process. According to

Castleton [10], the moisture levels of the soil influence the heat loss through evapotranspiration. Wet soils can promote heat removal from the building when evapotranspiration effects are considerable.

Different studies have assessed the evapotranspiration effect of succulents in extensive green roofs. According to Berghage *et al.* [27], it has been previously hypothesized that transpiration from succulents is not significant when compared to evaporation from bare substrate. However, these authors stated that the transpiration has a relevant role in the evapotranspiration process under well-watered soil conditions; and during dry conditions, these plants stop transpiring, leading to similar levels of evapotranspiration and evaporation [28].

Based on studies presented by Rezaei and Jarret [29], depending on the seasonal condition, respectively for winter and summer, planted plots with succulents transferred to the atmosphere 34% and 51% more water due to the evapotranspiration process when compared to bare soil. According to experiments performed by Voyde *et al.* [30], the observed results challenge the hypothesis that succulent plants hold water all the time. Furthermore, using a lightweight modular system, their results indicated that when the water supply is not limited, transpiration contributes approximately 48% of the evapotranspiration.

Regarding the influence of the water in the storage systems, Alcazar and Bass's [16] studies revealed that water storage can improve the thermal performance of green roofs. Even though the water levels have not been monitored, their results show a reduction in *U*-values compared to systems without water storage. Furthermore, no monitoring has been carried out in the present work, and the water levels do not seem to have an important role in green roof thermal performance.

Due to water's thermal mass, a delay would be expected in the Sydney and Rio de Janeiro results between temperature peaks considerably greater than the observed ones. Additionally, the modular system adopted herein is simply supported on top of the roof, and thus, it promotes the passage of air, which is believed to enhance the thermal insulation more so. It should be pointed out also that only temperature, and not solar radiation levels, were collected in the two exploratory studies presented. Thus, it is posited that the temperature attenuation provided by the green roofs must be directly related to high solar radiation levels and that during cloudy days, this effect tends to be less pronounced.

4. Conclusions

Both experimental setups in Rio de Janeiro and Sydney have demonstrated the potential for lightweight, portable green roof modules to be retrofitted on metal sheet roofing as a means of cooling buildings, reducing carbon emissions and contributing towards zero carbon targets. However, the exploratory experiments carried out in Sydney presented a potentially better green roof performance in temperature attenuation, which may be attributed to the closer positioning of the data loggers in relation to the roof.

There are different typologies regarding green roof technologies. At different scales, from extensive to intensive systems, it is expected that different soil depths, substrate composition, water retention layers, types of plants, soil moisture content, *etc.*, have a particular role in heat transmission, thermal inertia and evapotranspiration. The low cost, modular, portable, lightweight green roof adopted here comprises an extensive system, where all the components are gathered in a single module. The only difference between the experimental systems explored here and the conventional ones is that there is no full contact between the modules and the underlying roof, which provides some air circulation (and some extra insulating effect). However, considering that there is partial contact between the modules and the roof, it was, in fact, observed that there was some effect on inertia and the time shift when the inner temperatures of the bare and green roofs are compared. Similar results from experiments where modular systems were employed [27] show the same trend observed in the present exploratory study.

The *U*-value parameter has been used to quantify the heat gain or loss through a surface according to different works presented in the literature so far. In addition to Wong *et al.* [15], who evaluated this parameter from exposed roofs to covered roofs with different types of vegetation and soil substrates, Nichaou *et al.* [14], Alcazar and Bass [16] and Castleton [10] presented an evaluation of this parameter, especially for lightweight green roofs with depths of 75–100 mm and 50 mm, respectively. These works not only corroborate the role of the green roofs in reducing *U*-values substantially when compared to exposed roofs, but also noted the effects of evapotranspiration, thermal inertia, soil moisture content and water storage. According to Castleton [10], in addition to insulation, green roofs add thermal mass and provide inertia against temperature fluctuation. A green roof may work as a passive cooler. In the case of high external temperatures, not only the entering heat flux is cancelled, but also a slight outgoing flux is produced due to the cooling effect of the evapotranspiration.

The water levels in the soils are shown to affect the extent of heat loss due to the evapotranspiration process. The thermal conductivity of the soil is proportional to the water content, as drier soil conditions offer better thermal insulation.

An alternative to a green roof; known as a cool roof, is to paint the roof surface white, and this is common practice in the Mediterranean. This approach lowers the air and surface temperature, and less heat energy propagates into the building. According to Diep [31], cool or white roofs have an extreme capacity to reflect sunlight and heat. Green roofs do not reflect as well as white roofs. However, both roofs manage to cool buildings on hot days. During summer periods, green roofs have a major role in energy savings, due to the cooling effect of evapotranspiration, while in the winter, they are able to reduce heat loss, because of their more pronounced insulation properties.

Water levels were not monitored in this feasibility study. However, it is noted that due to the delay between the temperature peaks observed, the water levels do not seem to have a predominant role in green roof thermal performance. Furthermore, the passage of air is allowed in the modular system adopted, due to the existing space between the planted trays and the underside of the roof. This characteristic seems to enhance the insulating properties of the system and/or mitigate the effects of the water storage.

Even though no low temperatures (that is <16 °C) were measured, the negative differences observed (the green roof temperature was higher than the non-green roof) may indicate the potential for green roofs to attenuate extremes of temperature, due to their insulation properties. It is probable that different substrates would provide different results, and this should be investigated.

The temperature differences showed a relationship to the temperature background. The slight delay between temperature peaks of non-green and green roofs results in slightly warmer green roof temperatures (negative differences) during the nighttime and in the early morning periods, which contribute to weakening the temperature differences during the following day.

Considerable differences of temperature between city centres and suburban urban areas have been reported in the literature. Green roofs promote thermal comfort improvement, attenuating heat exchanges between the internal and external environments of buildings. Additionally, as these results suggest, attenuation of the urban heat island effect in large cities can be achieved, if green roofs are adopted for new building and as retrofitting for existing buildings on a city scale. For example, a modelling study for Toronto, Canada, predicted that adding green roofs to 50% of the available surfaces downtown would cool the entire city by 0.1 to 0.8 °C. Irrigating these roofs could further reduce temperatures by about 2 °C and extend a 0.5–1 °C cooled area over a larger geographic region [32]. The research has demonstrated that roof structures planted with succulent plants are viable, as the plants survived well and could provide a low-cost, drought-tolerant, lightweight option to reduce heat gain and heat loss through roof structures in some regions of NSW and Australia.

However, as far as the thermal effect is concerned, the adoption of green roofs in urban centres is a partial solution, due to the contribution of the building facades in the overall heating. Thus, a combination of green walls and green roofs could be an optimum solution for this problem. Furthermore, with regard to energy saving issues, considering that buildings comprise the most part of big cities, the use of green roofs would only bring an effect to top floors, which reinforces the combination of these systems (green roof and green walls) in the urban environments.

Additional experiments with structures that more closely emulate typical Australian housing specifications in terms of wall construction would be very useful to consolidate the preliminary results found here. One of the limitations of this exploratory research is that the walls of the shed are profile metal sheeting, which is not typically specified in housing, although it does exist. In the Rio de Janeiro experiments, brick walls comprise a common type of solution adopted in the majority of housing. However, additional procedures, such as green walls, should be evaluated in order to mitigate the existing thermal exchanges through the walls.

The lightweight modular green roof system adopted in our prototypes in Rio have been in testing for approximately 2.5 years, during which time, no maintenance has been required. The succulent plants have shown extreme resistance to intense heat, drought periods and atmospheric pollutants. The Sydney experiment has been underway for a year, and no maintenance of the plants has been required.

This study, as performed to date, used a modular system that has enabled new adaptive techniques to be adopted, where the green roof set-up costs have been substantially reduced using reusable materials. This novel approach is being tested at the Fiocruz campus using cellular plastic boxes, which are commercially available at a low cost. Additionally, these boxes are also being adopted as a new green wall system, arranged side by side in "U" metal profiles bolted on the walls. This new modular system has been shown to be applicable to roofs and walls and comprises a portable green patch solution that is intended for widespread use at a large scale in urban environments, which will lead to better environmental outcomes for the inhabitants.

Acknowledgments

Sara Wilkinson would like to thank the support of the Centre for Contemporary Design Practices (CCDP) within the Faculty of Design Architecture and Building at UTS for their support.

Renato Castiglia Feitosa would like to thank the support provided by The National Council for Scientific and Technological Development (CNPq), Brazil, Science without Borders program.

Author Contributions

Both authors contributed equally to this paper.

Conflicts of Interest

The authors declare no conflict of interest.

References

1. Australian Bureau of Meteorology (BOM). Available online: http://www.bom.gov.au (accessed on 18 March 2014).
2. Inter-governmental Panel Climate Change (IPCC). Climate Change 2013: The Physical Science Basis. Available online: http://www.ipcc.ch/report/ar5/wg1/ (accessed on 13 July 2014).
3. Maslin, M.; Austin, P.; Dickson, A.; Murlis, J.; Owen, M.; Panizzo, V. Audit of UK Greenhouse Gas emissions to 2020: Will current Government policies achieve significant reductions? Available online: http://www.ucl.ac.uk/environment-institute/research/dispatches (accessed on 21 March 2014).
4. Kelly, M.J. Britain's Building Stock—A Carbon Challenge 2008. Available online: http://www.lcmp.eng.cam.ac.uk/wp-content/uploads/081012_kelly.pdf (accessed on 14 March 2014).
5. Lamond, J.; Wilkinson, S.; Rose, C. Conceptualizing the benefits of green roof technology for commercial real estate owners and occupiers. In Proceedings of Pacific Rim Real Estate Conference, Christchurch, New Zealand, 19–22 January 2014.
6. Oke, T.R. The energetic basis of the urban heat island. *Q. J. Roy. Meteor. Soc.* **2006**, *108*, 1–24.
7. Harlan, S.; Brazer, A.J.; Prashad, L.; Stefanov, W.L.; Larsen, L. Neighborhood microclimates and vulnerability to heat stress. *Soc. Sci. Med.* **2006**, *63*, 2847–2863.
8. Climatempo. Available online: http://www.climatempo.com.br/destaques/tag/recorde-de-calor/ (accessed on 24 March 2014).
9. ABC, Heatwave Blamed for Large Spike in the Number of Deaths in Victoria Last Week. Available online: http://www.abc.net.au/news/2014-01-23/heatwave-death-toll-expected-to-top-almost-400/5214496 (accessed on 21 March 2014).
10. Castelton, H. Green roofs; building energy savings and the potential for retrofit. *Energ. Build.* **2010**, *42*, 1582–1591.
11. Wilkinson, S.J.; Reed, R. Green Roof Retrofit Potential in the Central Business District. *J. Prop. Manag.* **2009**, *27*, 284–301.

12. City of Sydney Greening Sydney Plan May 2012. Available online: http://www.cityofsydney.nsw.gov.au (accessed on 15 March 2014).
13. Williams, N.; Raynor, J.; Raynor, K. Green roofs for a wide brown land: Opportunities and barriers for rooftop greening in Australia. *Urban For. Urban Green.* **2010**, *9*, 169–272.
14. Niachou, A.; Papakonstantinou, K.; Santamouris, M.; Tsangrassoulis, A.; Mihalakakou, G. Analysis of the green roof thermal properties and investigation of its energy performance. *Energ. Build.* **2001**, *33*, 719–729.
15. Wong, N.H.; Cheong, D.K.W.; Yan, H.; Soh, J.; Ong, C.L.; Sia, A. The effects of rooftop garden on energy consumption of a commercial building in Singapore. *Energ. Build.* **2003**, *35*, 353–364.
16. Alcazar, S.; Bass, B. Energy performance of green roofs in a multi storey residential building in Madrid. In *Greening Rooftops for Sustainable Communities*; Green Roofs for Healthy Cities: Washington, DC, USA, 2005
17. Hes, D.; Livesley, S.; Jensen, S.; Aye, L.; Williams, N.; Cardorel, X. Healthy cities, climate change and the role of high albedo surfaces and green infrastructure. In Proceedings of the 5th Healthy Cities: Working Together to Achieve Liveable Cities Conference, Geelong, Australia, 6–8 June 2012.
18. Wilkinson, S.J.; Ghosh, S.; Page, L. Options for green roof retrofit and urban food production in the Sydney CBD. In Proceedings of the RICS COBRA Conference, New Delhi, India, 10–12 September 2013.
19. Kellert, S.R.; Wilson, E.O. *The Biophilia Hypothesis.* Island Press: Washington, DC, USA, 1993.
20. 202020 Vision. Available online: http://greenmagazine.com.au/202020-vision/ (accessed on 21 March 2014).
21. Agência Brasil. Available online: http://memoria.ebc.com.br/agenciabrasil/noticia/2012-06-02/area-verde-por-habitante-cai-26-no-rio-com-avanco-de-favelas-e-especulacao-imobiliaria (accessed on 15 March 2014).
22. City of Sydney. Green Roofs and Walls Policy. http://www.cityofsydney.nsw.gov.au/vision/towards-2030/sustainability/greening-the-city/green-roofs-and-walls (accessed on 10 January 2015).
23. Mentens, J.; Raes, D.; Hermy, M. Green roofs as a tool for solving the rainwater runoff problem in the urbanized 21st century? *Landsc. Urban Plan.* **2006**, *77*, 217–226.
24. Feitosa, R.C. Telhados Verdes Reduzem Temperatura Interna de Casas. Globo Ciência 2012. Available online: http://oglobo.globo.com/ciencia/telhados-verdes-reduzem-temperatura-interna-de-casas-5173467 (accessed on 10 March 2014).
25. Del Barrio, E.P. Analysis of the green roofs cooling potential in buildings. *Energ. Build.* **1998**, *27*, 179–193.
26. Lazzarin, R.M.; Castellotti, F.; Busato, F. Experimental measurements and numerical modelling of a green roof. *Energ. Buildings* **2005**, *37*, 1260–1267.

27. Berghage, R.D.; Jarrett, A.R. *Quantifying Evaporation and Transpirational Water Losses from Green Roofs and Green Roof Media Capacity for Neutralizing Acid Rain*; National Decentralized Water Resources Capacity Development Project; Penn State University: State College, PA, USA, 2007.
28. Lin, Y.-J.; Lin, H.-T. Thermal performance of different planting substrates and irrigation frequencies in extensive tropical rooftop greeneries. *Build. Environ.* **2011**, *46*, 345–355.
29. Rezaei, F.; Jarrett, A.R.; Berghage, R.D.; Beattie, D.J. Evapotranspiration Rates from Extensive Green Roof Plant Species. In Proceedings of the 2005 ASAE Annual International Meeting Sponsored by ASAE, Tampa, FL, USA, 17–20 July 2005.
30. Voyde, E.; Fassman, E.; Simcock, R.; Wells, J. Quantifying Evapotranspiration Rates for New Zealand Green Roofs. *J. Hydrol. Eng.* **2010**, *15*, 395–403.
31. Diep, F. White Roofs Keep Cities Cooler Than "Green" Ones, Study Find. Available online: http://www.popsci.com/article/science/white-roofs-keep-cities-cooler-green-ones-study-finds (accessed on 27 January 2014).
32. EPA, 2014. Reducing Urban Heat Islands: Compendium of Strategies. Green Roofs. Available online: http://www.epa.gov/heatislands/resources/pdf/GreenRoofsCompendium.pdf (accessed on 2 November 2014).

The Effects of Air Permeability, Background Ventilation and Lifestyle on Energy Performance, Indoor Air Quality and Risk of Condensation in Domestic Buildings

Arman Hashemi and Narguess Khatami

Abstract: Effective and efficient ventilation is essential when improving energy performance and Indoor Air Quality (IAQ) of buildings. Reducing air permeability can considerably improve the energy performance of buildings; however, making the buildings more airtight may result in lower rates of natural ventilation which may in turn increase the risks of condensation and unacceptable IAQ. This study evaluates the effects of different air permeability rates, background ventilation and occupants' lifestyles on the energy performance as well as the risk of condensation and CO_2 concentration in domestic buildings. Dynamic computer simulations were conducted in EnergyPlus. Results indicated direct relations between the ventilation rates, energy performance and IAQ. Higher air permeability along with background ventilation resulted in considerably better IAQ while energy consumption increased by up to four times. Occupants' lifestyles were identified as a major contributor to the risk of condensation.

Reprinted from *Sustainability*. Cite as: Hashemi, A.; Khatami, N. The Effects of Air Permeability, Background Ventilation and Lifestyle on Energy Performance, Indoor Air Quality and Risk of Condensation in Domestic Buildings. *Sustainability* **2015**, 7, 4022-4034.

1. Introduction

Britain has the oldest housing stock in the developed world [1]. According to BRE [2] more than 65% and 49% of domestic buildings in the UK have uninsulated walls and single glazed windows respectively. It is estimated that domestic buildings are responsible for one third of the UK carbon emissions [3,4]. Improving the quality and energy performance of new and existing buildings has therefore been one of the major priorities of the UK Government during the recent years. This is not only to achieve the CO_2 emission targets set for 2050 [5,6] but also to tackle fuel poverty which affects more than 2.3 million households [7]. National schemes such as the Green Deal [8] and Warm Front intend to improve energy efficiency of buildings through better insulation and reduced air permeability to reduce the risk of thermal discomfort as well as unnecessary heat loss [9]. Building regulations and standards have also been increasing thermal insulation and airtightness requirements to improve the energy performance of new and refurbished buildings in the UK.

According to Xing *et al.* [10] improving airtightness is an effective way of reducing energy consumption in buildings. Yet, high airtightness could result in insufficient ventilation rates which may in turn lead to poor indoor air quality (IAQ), high risk of condensation and reduced comfort and productivity [11,12]. According to the Scottish Government [13], when refurbishing existing properties, heating, ventilation and thermal insulation should be considered together as ignoring one could result in poor indoor conditions, higher risk of condensation and unnecessary energy consumption.

The results of a study by Staepels *et al.* [14] suggest that increasing airtightness in new buildings would not necessarily deteriorate the IAQ as Relative Humidity (RH%) and CO_2 levels may remain within the acceptable ranges. On the contrary, results of another study by of Hobday [15] reveal that although increasing airtightness helps to reduce energy consumption and risk of thermal discomfort, it could considerably increase the risk of poor IAQ. Boulanger *et al.* [16] also reported considerably higher RH% levels in airtight buildings. Moreover, results of studies by Elsharkawy and Rutherford [17] indicate that in energy efficient buildings, the occupants' lifestyle does not change as fast as the other parameters and therefore, in actual buildings both energy consumption and IAQ are worse than the predicted conditions.

For very low air permeability rates, such as 1 $m^3/h.m^2$@50 pa in Zero Carbon Homes, mechanical ventilation systems are required [18,19] to maintain an acceptable IAQ in the building in order to avoid negative effects on the health and wellbeing of the occupants. Nevertheless, the results of a study by Wing and Huang [20] show that applying mechanical ventilation in domestic buildings increased the number of sick building syndrome incidents and affected occupants' perceived comfort. Although the situation slightly improved for buildings where fresh air was supplied by means of mechanical ventilation, CO_2 levels and occupants comfort remained worse than naturally ventilated buildings.

It should be noted that application of mechanical ventilation in UK domestic buildings is not a common practice and natural ventilation remains the main strategy for the majority of new and refurbished houses in the UK. Reducing air permeability could therefore result in higher risk of condensation, and unacceptable IAQ, affecting the health and wellbeing of the occupants of these buildings. To this end, this study intends to investigate the effects of air permeability, background ventilation and occupants' lifestyles on the energy performance, indoor air quality and risk of condensation in naturally ventilated domestic buildings. The ultimate aim is to determine the most efficient way to improve energy efficiency in buildings without sacrificing IAQ.

2. Methodology

Dynamic Thermal Simulations (DTS) were conducted in EnergyPlus to evaluate building performance. Combined Heat and Moisture Finite Element was considered as the heat balance algorithm in order to achieve more accurate results for simulating moisture and the risk of condensation. London Gatwick was considered for the weather data.

Simulations were conducted for the heating seasons from October to the end of March when occupants are less likely to open the windows/vents and there are therefore higher risks of condensation and poor IAQ. According to the findings of a study by Hashemi *et al.* [21] on some UK domestic buildings, bedrooms have a considerably poorer IAQ compared to other areas of the building. Therefore, for the purpose of this study, a 3 m × 3 m × 2.4 m (length × width × height) room with a 1 m × 1 m window and masonry insulated cavity walls was modelled as the representative of a double bedroom.

Two occupants were defined as the only sources of internal heat gain and CO_2 generation in the simulated zone. According to ASHRAE 55 [22] total heat gain of 120 W was assumed for seated and 72 W for sleeping occupants. The occupancy schedule was defined as fully occupied from 6 pm to 8 am during weekdays and 24 h during all other days. The bedrooms were supposed to be occupied for 8 h during working days and 9 h during all other days. An average heating set-point of 18 °C was considered during the occupied periods. This was specified based on the CIBSE Guide A [23] guidelines in which temperature of 17 °C–19 °C is suggested for bedrooms.

In total, 30 different scenarios were simulated for five airtightness rates (1, 3, 5, 7 and 10 $m^3/h.m^2$@50 Pa) with two Trickle Vents (TV) positions (permanently closed or open) and three different lifestyle conditions of (Dry, Moist and Wet). It should be noted that according to approved document L1A, the limiting air permeability in domestic buildings is 10 $m^3/h.m^2$@50 Pa [24]. CIBSE Technical Memoranda TM23: 2000 also recommends an air permeability of 5 m^3/h.m2@50 Pa as the best practice for naturally ventilated dwellings [25]. For the purpose of background/whole building ventilation a TV of 5000 mm^2 was considered based on the approved document part F of the UK building regulations [26]. Background ventilation is necessary to maintain the minimum IAQ requirements in the building. The ventilation was supposed to be through infiltration and TV only and no obstructions such as curtains or blinds were considered.

Regarding moisture generation, three lifestyle categories of "Dry", "Moist" and "Wet" occupancy conditions were modelled based on the guidelines provided in Annex B, Table B.3 of BS 5250 [27]. According to the provided guidelines, Dry, Moist and Wet occupancy conditions are defined as excess generated moisture of up to 0.3 kPa, 0.6 kPa and more than 0.6 kPa, respectively. For the Moist and Wet occupancy, internal humidity is higher than normal levels, while dry occupancy is representative of those conditions where occupants are out of the building during the daytime and generate lower amount of moisture. Therefore, for the purpose of simulations, excess vapour pressure values of 0.3 kPa, 0.6 kPa and 0.9 kPa were assumed as the Dry, Moist and Wet occupancy conditions, respectively. According to the BS 5250, the daily moisture generation of, for example, two occupants for Dry, Moist and Wet occupancy conditions are 4 kg, 8 kg and 11 kg, respectively. Since window surface temperatures are usually lower than the rest of the building elements, risk of condensation and mould growth around windows is usually higher than in other areas of buildings. The total number of hours when there was a risk of condensation on the window are therefore reported.

Moreover, as suggested by BB101 [28] and BS EN 15251 [29], CO_2 concentration can be used as an indicator of IAQ in occupied spaces. High CO_2 levels in occupied zones is an indicator of unacceptable IAQ and insufficient ventilation. The indoor air quality was therefore investigated by reporting CO_2 concentration level during occupied periods. For this reason, frequency of the times when CO_2 levels were higher than 1000, 1500 and 5000 parts per million (ppm) are reported. The 1000 ppm value is the target value suggested by BS EN 15251:2007 [29] as the acceptable IAQ while 1500 ppm is the concentration value in which usually all the occupants show some symptom of sick building syndrome [30] and the 5000 ppm is the concentration value in which occupants' health can be affected [31].

In addition, the effects of the application of each of 30 scenarios on energy consumption were also studied and the total heating energy consumptions were reported. It should be noted that the reported heating energy is for the studied thermal zone only and the results should either be divided by the total area or the volume of the heated zone in order to achieve heating energy per square/cubic meters as a more appropriate indicator for comparison with other buildings. Table 1 summarises the conditions considered for the simulations.

Table 1. Summary of conditions/scenarios considered for the simulations.

Summary of Conditions					
Simulation Conditions	Simulation period: 6 Months (October-March)	Weather Data: London Gatwick	Heating Set-point: 18 °C	Occupancy pattern: 6 pm–8 am working days; Occupied all other times	Number of occupants: 2
Construction	Masonry cavity with insulated walls	External walls U-Value: 0.35 $W/m^2 \cdot K$	Window U-Value: 1.97 W/m^2K	Room size: 3 m × 3 m × 2.4 m	-
Boundary conditions	Exposed: South wall; Roof	Ground: Floor	Adiabatic: North, East and West walls	Total exposed wall area: 7.2 m^2	-
Air Permeability Rates	1 $m^3/h \cdot m^2$@50 Pa	3 $m^3/h.m^2$@50 Pa	5 $m^3/h \cdot m^2$@50 Pa	7 $m^3/h.m^2$@50 Pa	10 $m^3/h \cdot m^2$@50 Pa
Occupancy Conditions	Dry: 0.3 Kpa excess vapour pressure	Moist: 0.6Kpa excess vapour pressure	Wet: 0.9 Kpa excess vapour pressure	-	-
TV Conditions	Open	Closed	Area: 5000 mm^2	Position: South façade	Discharge coefficient: 0.65
Simulation Outputs	CO_2: 1000 ppm, 1500 ppm, 5000 ppm	Risk of Condensation (h)	Energy Consumption (KW)	-	-

3. Results

3.1. Indoor Air Quality

The results of simulations were evaluated for different airtightness values for the TV open and closed conditions. According to the results for the air permeability of 1 $m^3/h.m^2$@50 Pa when the TV was closed, CO_2 concentration levels were above 1000 ppm, 1500 ppm and 5000 ppm for 85%, 69% and 10% of all occupied periods, respectively. The IAQ improved considerably when the TV was opened, reducing the above figures to 54%, 31% and 1.5% of the entire occupied periods. Opening the TV helped to reduce risk of poor IAQ by 85% reflecting the period of time when CO_2 level was higher than 5000 ppm.

For the 3 $m^3/h.m^2$@50 Pa air permeability, the results indicated considerable reduction to around 31% during which CO_2 was above 1500 ppm. However CO_2 remained above 1000 ppm in 54% of the occupied periods. The IAQ continued to improve for all other air permeability values. For the air permeability of 10 $m^3/h.m^2$@50 Pa, CO_2 concentration was respectively in 8.3% and 1% of the times above 1500 ppm and 5000 ppm. This means 79%, 88% and 90% improvements in the

CO_2 concentration for 1000, 1500 and 5000 ppm compared to 1 $m^3/h.m^2$@50 Pa. Figure 1 and Table 2 summarise the results of simulations for different air permeability values when the TV was closed.

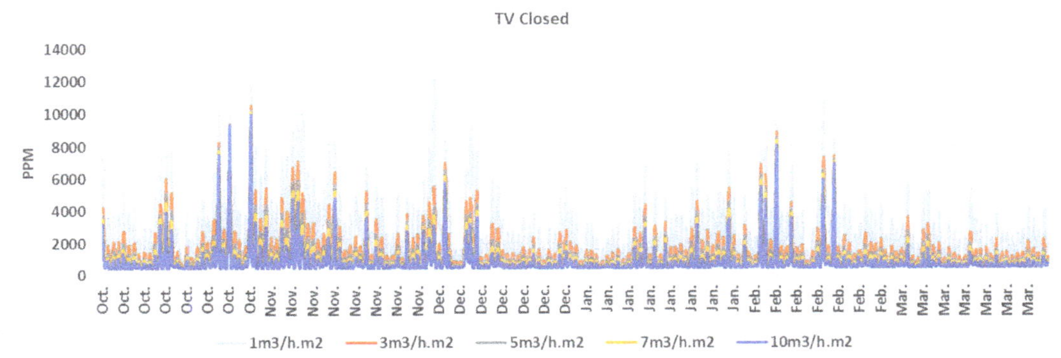

Figure 1. CO_2 concentration values for different air permeability rates Oct.–Mar. (TV Closed).

Table 2. Percentage of occupied time when CO_2 concentration was above, 1000, 1500 and 5000 ppm (TV Closed).

Air Permeability Values (TV Closed) / CO_2 Concentration	1$m^3/h.m^2$@ 50 Pa	3$m^3/h.m^2$@ 50 Pa	5$m^3/h.m^2$@ 50 Pa	7$m^3/h.m^2$@ 50 Pa	10$m^3/h.m^2$@ 50 Pa
1000 ppm	85.3%	54.3%	36.3%	25.8%	17.8%
1500 ppm	69.6%	31.3%	17.7%	12.0%	8.3%
5000 ppm	10.3%	2.3%	1.5%	1.3%	1.0%

Opening the TV also significantly improved the IAQ. The best conditions in terms of IAQ was, as expected, achieved for the air permeability of 10 $m^3/h.m^2$@50 Pa. For this condition, CO_2 concentration was above 1000 ppm and 1500 ppm for 14.5% and 6.8% of the occupied periods, respectively. The CO_2 concentration was above 5000 ppm for 0.6% of the times. This implies a 73%, 78% and 60% improvement in the CO_2 concentration over the 1 $m^3/h.m^2$@50 Pa air permeability case for 1000, 1500 and 5000 ppm, respectively. Figure 2 and Table 3 summarise the results of simulations for different air permeability values when TV was open.

According to the results, increasing the air permeability by 2 $m^3/h.m^2$@50 Pa and/or opening the TV had almost identical effects on the CO_2 concentration values. For instance a 3 $m^3/h.m^2$@50 Pa air permeability with open TV would result in almost the same CO_2 concentration levels for the air permeability of 5 $m^3/h.m^2$@50 Pa with closed TV.

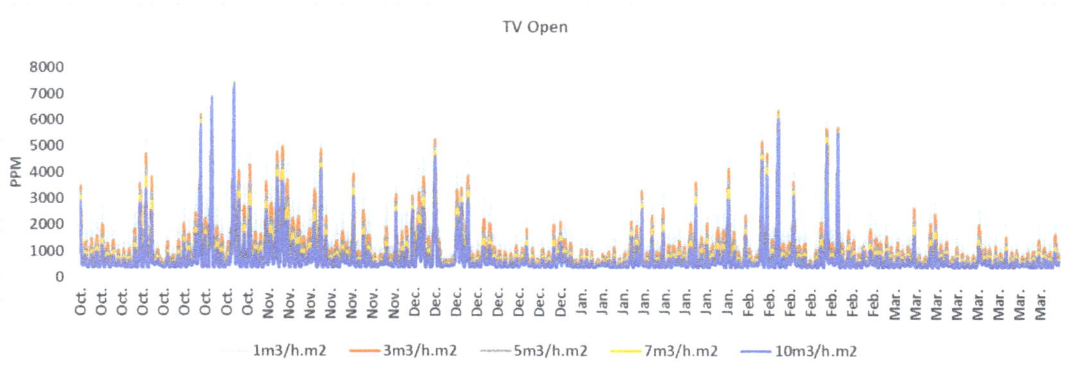

Figure 2. CO_2 concentration values for different air permeability rates Oct.-Mar. (TV Open).

Table 3. Percentage of occupied time when CO_2 concentration was above, 1000, 1500 and 5000 ppm (TV Open).

Air Permeability Values (TV Open) / CO_2 Concentration	1 $m^3/h.m^2$@ 50 Pa	3 $m^3/h.m^2$@50 Pa	5 $m^3/h.m^2$@ 50 Pa	7 $m^3/h.m^2$@ 50 Pa	10 $m^3/h.m^2$@ 50 Pa
1000 ppm	54.3%	36.7%	25.8%	19.4%	14.5%
1500 ppm	31.1%	17.5%	11.7%	8.7%	6.8%
5000 ppm	1.5%	1.0%	0.7%	0.6%	0.6%

3.2. Risk of Condensation

This section evaluates the effects of different occupancy conditions/lifestyles ("Dry", "Moist" and "Wet" occupancy) on the risk of condensation. The results show strong relations between the risk of condensation and lifestyle as an indicator of generated moisture by the occupants during the day. According to the results, there is a very high risk of condensation for all Wet occupancy conditions regardless of the air permeability and/or the position of the TV. The situation moderately improved for the Moist and considerably improved for the Dry occupancy conditions. Tables 4 and 5 summarise the risk of condensation for different air permeability values and lifestyle conditions with and without TV.

The risk of condensation was negligible for almost all Dry occupancy conditions except for the 1 $m^3/h.m^2$@50 Pa air permeability during October when 159 h of condensation was observed when the TV was closed. Higher air permeability did not generally improve the situation although it greatly improved the conditions in October for the 1 $m^3/h.m^2$@50 Pa air permeability. Similar effects were observed for the rest of airtightness rates when the TV was opened. The results also indicate that, apart from October, opening the TV did not improve the conditions, and marginally deteriorated the situations resulting in slightly higher risk of condensation. A possible explanation for this may be that opening the TV would reduce the surface temperature of the window and surrounding walls resulting in higher risk of condensation. More research is required in this respect.

Table 4. Number of hours when there was a risk of condensation for different air permeability rates and occupancy conditions (TV Closed).

TV Closed	1 m³/h.m²@ 50 Pa	3 m³/h.m²@ 50 Pa	5 m³/h.m²@ 50 Pa	7 m³/h.m²@50 Pa	10 m³/h.m²@ 50 Pa
			Dry Occupancy		
October	159.1	5.7	0.0	0.0	0.0
November	40.2	0.0	0.1	0.1	0.3
December	9.2	3.1	3.5	4.0	5.2
January	0.4	0.9	1.5	2.2	3.5
February	3.2	6.5	8.6	9.9	11.9
March	0.2	0.4	0.8	1.3	1.7
Total (Hours)	212.3	16.6	14.5	17.5	22.6
			Moist Occupancy		
October	161.4	21.8	12.6	18.5	24.0
November	71.4	80.4	105.6	117.8	125.6
December	211.9	222.5	229.9	235.6	241.3
January	286.3	323.8	331.6	335.1	338.7
February	232.8	250.0	264.5	271.3	277.1
March	72.6	117.6	139.3	149.7	161.4
Total (Hours)	1036.4	1016.1	1083.5	1128	1168.1
			Wet Occupancy		
October	206.9	133.8	185.7	224.3	257.1
November	305.7	357.7	393.3	411.8	422.1
December	634.2	593.7	569.1	558.4	546.2
January	613.7	580.2	561.9	549.2	524.4
February	556.2	517.9	506.6	489.7	478.9
March	479.2	477.7	474.0	472.2	470.5
Total (Hours)	2795.9	2661	2690.6	2705.6	2699.2

Table 5. Number of hours when there was a risk of condensation for different air permeability rates and occupancy conditions (TV Open).

TV Open	1 m³/h.m²@ 50 Pa	3 m³/h.m²@ 50 Pa	5 m³/h.m²@ 50 Pa	7 m³/h.m²@ 50 Pa	10 m³/h.m²@ 50 Pa
			Dry Occupancy		
October	3.5	0.0	0.0	0.0	0.0
November	0.0	0.1	0.1	0.3	0.5
December	2.4	3.3	4.2	5.0	6.1
January	0.9	1.5	2.2	3.1	4.1
February	6.5	8.5	9.9	11.4	13.0
March	0.3	0.8	1.4	1.6	1.9
Total (Hours)	13.6	14.2	17.8	21.4	25.6

Table 5. *Cont.*

TV Open	1 m³/h.m²@ 50 Pa	3 m³/h.m²@ 50 Pa	5 m³/h.m²@ 50 Pa	7 m³/h.m²@ 50 Pa	10 m³/h.m²@ 50 Pa
			Moist Occupancy		
October	12.0	11.4	14.3	18.5	22.6
November	78.8	104.4	117.5	122.9	129.3
December	221.4	230.1	234.9	240.2	243.8
January	324.8	331.7	335.3	337.7	340.0
February	250.1	264.1	270.6	275.0	279.8
March	115.5	138.6	150.0	158.1	166.6
Total (Hours)	1002.6	1080.3	1122.6	1152.4	1182.1
			Wet Occupancy		
October	133.7	187.1	225.9	248.5	271.3
November	356.4	395.3	412.4	419.8	427.4
December	597.6	570.4	559.0	551.8	543.0
January	580.3	562.4	549.2	533.0	514.3
February	518.1	506.9	489.9	481.9	474.4
March	476.9	474.0	472.0	471.2	469.5
Total (Hours)	2663	2696.1	2708.4	2706.2	2699.9

Natural and background ventilation due to the higher air permeability and/or through TVs, seem to be adequate for the dry occupancy conditions only. Wet occupancy conditions increased the risk of condensation by up to 2.7 times compared to the Moist conditions. Yet, none of the Moist and Wet occupancy conditions achieved desirable results as very high risk of condensation was observed for these conditions compared to the Dry occupancy condition. Either mechanical or purge ventilation through windows/vents may be required to mitigate the risk of condensation in these conditions.

3.3. Energy Consumption

Energy consumption was also investigated for all of the scenarios. The results of the simulations reveal direct relations between energy consumption and ventilation rates in general, and air permeability and background ventilation in particular. According to the results, energy consumption during heating seasons increased by around thirteen times (from 7.5 kW/m^2 to 96.6 kW/m^2) for the closed TV condition when the air permeability increased from 1 to 10 m^3/h.m^2@50 Pa (such an increase in energy consumption may not occur in real world conditions). For the open TV, energy consumption increased by around four times. It should be noted that the background ventilation is necessary to achieve acceptable IAQ in airtight buildings, and therefore, this fourfold increase in energy consumption may be a better reflection of the real world conditions. Figure 3 summarises the energy performance of the building from October to March for different air permeability ranges when the TV was open or closed.

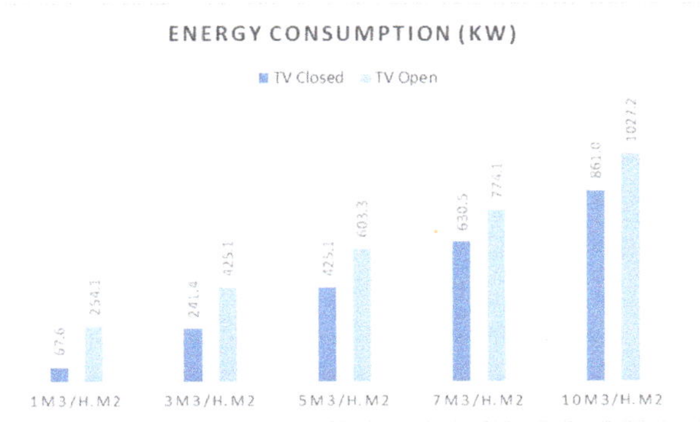

Figure 3. Total energy consumption between October and March for different air permeability rates when TV was open or closed.

Tables 6 and 7 also compare the relative increases in energy consumption for different scenarios during the heating seasons. According to the results of simulations for the TV open conditions, increasing air permeability by 2 $m^3/h.m^2$@50 Pa could increase the energy consumptions between 29%–42%. In scenarios where TV was closed, this figure was between 37%–76% excluding the 1 $m^3/h.m^2$@50 Pa air permeability (where there results are unrealistically low). The energy consumption increased by 2.4 times when comparing 3 $m^3/h.m^2$@50 Pa with 10 $m^3/h.m^2$@50 Pa for the TV open conditions. This figure was around 4 times when comparing 1 $m^3/h.m^2$@50 Pa with 10 $m^3/h.m^2$@50 Pa. The energy consumption increased by 3.6 times when comparing 3 $m^3/h.m^2$@50 Pa with 10 $m^3/h.m^2$@50 Pa for the TV closed conditions.

Table 6. Total energy consumption for different air permeability rates (TV Closed).

Air Permeability Values (TV Closed)	1 $m^3/h.m^2$@ 50 Pa	3 $m^3/h.m^2$@ 50 Pa	5 $m^3/h.m^2$@ 50 Pa	7 $m^3/h.m^2$@ 50 Pa	10 $m^3/h.m^2$@ 50 Pa
Energy Consumption					
Annual (KW)	68.9	265.2	493.6	713.8	1044.2
Heating Seasons (KW)	67.6	241.4	425.1	630.5	861.0
Relative Increase %	100.0%	357.3%	629.2%	933.2%	1274.3%
Relative Increase Compared to the previous %	N/A	100%	176.1%	148.3%	136.6%

Table 7. Total energy consumption for different air permeability rates (TV Open).

Air Permeability Values (TV Open)	1 $m^3/h.m^2$@ 50 Pa	3 $m^3/h.m^2$@ 50 Pa	5 $m^3/h.m^2$@ 50 Pa	7 $m^3/h.m^2$@ 50 Pa	10 $m^3/h.m^2$@ 50 Pa
Energy Consumption					
Annual (KW)	269.2	495.0	718.6	934.0	1249.9
Heating Seasons (KW)	254.1	425.1	603.3	774.1	1027.2
Relative Increase %	100.0%	167.3%	237.4%	304.7%	404.3%
Relative Increase Compared to the previous %	N/A	100%	141.9%	128.5%	132.7%

4. Conclusions

Improving the energy performance of new and existing buildings has been one of the major priorities of the UK Government during the recent years. National schemes such as the Green Deal and Warm Front intend to improve energy efficiency of buildings through better insulation and reduced air permeability. However, improving building energy performance without considering its effects on the other factors such as indoor air quality may negatively affect the health and wellbeing of the occupants. This paper demonstrated the effects of different air permeability rates and background ventilation on the energy performance, indoor air quality (IAQ) and risk of condensation in domestic buildings.

The results of this study reveal invers relations between energy performance and indoor air quality/ventilation rates in buildings. Higher ventilation rates by means of infiltration and/or background ventilation through, for example, Trickle Vents (TV) significantly decreased the risk of unacceptable IAQ; however, they increased the energy consumption by up to 4 times. Although very low air permeability rates significantly reduce energy consumption, they may not be capable of providing acceptable IAO if they rely on background ventilation alone. Larger ventilation areas and/or mechanical ventilation may therefore be required to achieve acceptable indoor air quality in airtight buildings.

The results of this study also reveal that occupants' behaviours/occupancy conditions are important factors in mitigating or increasing the risk of condensation in naturally ventilated buildings. Risk of condensation increased by up to 2.7 times for Wet occupancy conditions compared to Moist occupancy. Nevertheless, neither higher air permeability nor background ventilation through TVs are effective in controlling condensation in buildings with Moist and/or Wet occupancy conditions. However, for the Dry occupancy conditions, where the excess vapour pressure is not more than 0.3 kPa of the internal vapour pressure, all air permeability rates along with the background ventilation through TVs can achieve acceptable results in terms of the risk of condensation. For higher moisture generation rates, purge or mechanical ventilation may be required to effectively remove the excess moisture at the source. This said, it should be noted that other factors such as poor ventilation, design and detailing can also contribute to high risks of condensation, damp and mould growth in new and refurbished buildings.

Based on the results of this study, it could be argued that for naturally ventilated homes with Dry occupancy conditions, an air permeability of 3 $m^3/h.m^2$@50 Pa could achieve optimum results in terms of energy consumption and IAQ if adequate background ventilation is provided to achieve acceptable indoor air quality (e.g., CO_2 concentrations less than 1% of the occupied times above 5000 ppm). For the buildings where TV or other means of background ventilation is not provided, a minimum air permeability of 10 $m^3/h.m^2$@50 Pa is recommended to achieve acceptable IAQ. It should be noted that these figures may apply to households with a maximum of two occupants. For bigger households, higher air permeability rates and/or background ventilation may be required to achieve acceptable results.

Providing more airtight buildings along with controllable background ventilation may improve occupants' perceived thermal comfort by giving them more control over their environments.

However as people are usually unaware of CO_2 levels in the environment, [32] they may keep the TVs closed which may in turn increase the risk of poor indoor air quality particularly during colder seasons when occupants are less likely to open the windows/vents. Moreover, in view of the high costs of achieving airtight buildings, as far as energy performance and IAQ are concerned, it may be more economical to consider higher air permeability rates instead of increasing airtightness along with more background ventilation to achieve acceptable IAQ in naturally ventilated buildings.

In summary, the results of this study show that increasing airtightness may escalate the risk of poor IAQ by reducing the rate of ventilation; however, this relation was not as evident for the risks of condensation. The latter requires more investigation. The current building regulations and standards are less concerned about the IAQ in general and generated moisture in particular (as the most critical pollutant in domestic buildings [33,34]). It is therefore recommended to include detailed CO_2 and moisture control strategies/requirements in the building regulations and standards, as well as in the national energy efficiency schemes, not only to improve the energy performance of buildings but also to improve the health and wellbeing of the occupants of new and refurbished buildings. It is also recommended to refurbish buildings by carefully considering their current performance and evaluating the effects of refurbishment strategies on their future performance in terms of changes in risks of condensation and poor IAQ. Moreover, it should be noted that people's behaviours and lifestyles may not change as quickly as the building regulations are. Any modifications in the building regulations and standards should therefore take into account the effects of occupancy conditions and occupants' behaviours in order to avoid negatively affecting their health and wellbeing.

Author Contributions

The authors contributed equally to this work.

Conflicts of Interest

The authors declare no conflict of interest.

References

1. EST. *Energy-Efficient Refurbishment of Existing Housing*; Energy Saving Trust: London, UK, 2007.
2. BRE. *Energy Use in Homes: A Series of Reports on Domestic Energy Use in England. Energy Summary Report*; Building Research Establishment: Watford, UK, 2005.
3. English Heritage. *Energy Efficiency and Historic Buildings, Application of Part L of the Building Regulations to Historic and Traditionally Constructed Buildings*; English Heritage: Birmingham, UK, 2012.
4. DECC. *Energy Consumption in the UK (2013)*; Department for Energy and Climate Change: London, UK, 2013.
5. DECC. *Climate Act Impact 2008 Assessment*; Department of Energy & Climate Change: London, UK, 2009.

6. TSB. *Retrofit Revealed, The Retrofit for the Future Projects—Data Analysis Report*; Technology Strategy Board: Swindon, UK, 2013.
7. Department of Energy and Climate Change (DECC). *Annual Fuel Poverty Statistics Reports*; DECC: London, UK, 2014.
8. Department of Energy and Climate Change (DECC). *The Green Deal: A Summary of the Government's Proposals*; DECC: London, UK, 2010.
9. Hong, S.H.; Ridley, I.; Oreszczyn, T. The impact of energy efficient refurbishment on the airtightness in English dwellings. In Proceedings of the 25th AIVC Conference "Ventilation and retrofitting", Prague, Czech Republic, 15–17 September 2004.
10. Xing, Y.; Hewitt, N.; Griffths, P. Zero carbon buildings refurbishment—A Hierarchical pathway. *Renew. Sustain. Energy Rev.* **2011**, *15*, 3229–3236.
11. Bakó-Biró, Z.; Clements-Croome, D.J.; Kochhar, N.; Awbi, H.B.; Williams, M.J. Ventilation rates in schools and pupils' performance. *Build. Environ.* **2012**, *48*, 215–223.
12. Seppanen, O.; Fisk, W.J.; Lei, Q.H. Ventilation and performance in office work. *Indoor Air* **2005**, *6*, 28–36.
13. The Scottish Government. *Technical Handbook—Domestic*; The Scottish Government: Scotland, UK, 2013.
14. Staepels, L.; Verbeeck G.; Roels S.; van Gelder L.; Bauwens G. Do Ventilation Systems Accomplish the Necessary Indoor Air Quality in Low Energy Houses? In Proceedings of the 11th REHVA World Congress & 8th International Conference on IAQVEC, Prague, Czech Republic, 16–19 June 2013.
15. Hobday, R.A. *Indoor Environmental Quality in Refurbishment. Historic Scotland Research Report 12*; Historic Scotland: Edinburgh, Scottland, UK, 2011.
16. Boulanger, X.; Mouradian, L.; Pele, C.; Pamart, P.; Bernard, N. Lessons learned on ventilation systems from the IAQ calculations on tight energy performant buildings. In *Optimising Ventilative Cooling and Airtightness for [Nearly] Zero-Energy Buildings, IAQ and Comfort*, Proceedings of the 33rd Air Infiltration and Ventilation Centre, Copenhagen, Denmark, 11 October 2012.
17. Elsharkawy, H.; Rutherford, P. Retrofitting social housing in the UK: Home energy use and performance in a pre-Community Energy Saving Programme (CESP). *Energy Build.* **2015**, *88*, 25–33.
18. Department for Communities and Local Government (DCLG). *Code for Sustainable Homes, Impact Assessment*; DCLG: London, UK, 2009.
19. Laverge, J.; van den Bossche, N.; Heijmans, N.; Janssens, A. Energy saving potential and repercussions on indoor air quality of demand controlled residential ventilation strategies. *Build. Environ.* **2011**, *46*, 1497–1503.
20. Wing, N.H.; Huang, B. Comparative study of the indoor air quality of naturally ventilated and air-conditioned bedrooms of residential buildings in Singapore. *Build. Environ.* **2004**, *39*, 1115–1123.

21. Hashemi, A.; Liyanage, C.; Hadjri, K.; Khatami, N. Assessing the risks of dampness and mould growth in renovated properties. In Proceedings of the ZEMCH2014 International Conference: Mass Customisation and Sustainability in Housing, Londrina, Brazil, 4–6 June 2014.
22. ASHRAE. *Thermal Environmental Conditions for Human Occupancy*; American Society of Heating, Refrigerating, and Air-Conditioning Engineers (ASHRAE): Atlanta, GA, USA, 2004.
23. Chartered Institution of Building Services Engineers (CIBSE). *Guide A: Environmental Design*; CIBSE: London, UK, 2006.
24. HM Government. *The Building Regulations 2010 (ed 2013), Approved Document L1 A, Conservation of Fuel and Power in New Dwellings*; HM Government: London, UK, 2013.
25. CIBSE. *Technical Memoranda TM23: Testing Buildings for air Leakage*; Chartered Institution of Building Services Engineers (CIBSE): London, UK, 2000.
26. HM Government. *The Building Regulations 2010, Approved document F. Ventilation*; HM Government: London, UK, 2010.
27. BSI. Code of Practice for Control of Condensation in Buildings. In *BS EN 5250:2002*; British Standards Institute: London, UK, 2002.
28. DFE. *Ventilation of School Buildings: Regulations, Standards Design Guidance. Building Bulletin 101*; Department for Education and Employment: London, UK, 2006.
29. BSI. Indoor environmental input parameters for design and assessment of energy performance of buildings addressing indoor air quality, thermal environment, lighting and acoustics. In *BS EN 15251:2007*; British Standards: London, UK, 2007.
30. Petty, S.E. *Forensic Engineering: Damage Assessments for Residential and Commercial Structures*; Taylor & Francis: Boca Raton, FL, USA, 2013.
31. Emmerich, S.J.; Persily, A.K. *State-of-the-Art Review of CO_2 Demand Controlled Ventilation Technology and Application*; National Technical Information Service (NTIS): Springfield, IL, USA, 2001.
32. Khatami, N.; Cook, M.J.; Hudleston, N. Opportunities for Improving Environmental Performance of Retrofitted Lightweight Offices in Temperate Climates. In Proceedings of the ROOMVENT 2011: The 12th International Conference on Air Distribution in room, Trondheim, Norway, 19–22 June 2011.
33. EST. *Improving Airtightness in Dwellings*; Energy Saving Trust: London, UK, 2005.
34. EST. *Energy Efficient Ventilation in Dwellings—A Guide for Specifiers: GPG268*; Energy Saving Trust: London, UK, 2006.

Adopting Product Modularity in House Building to Support Mass Customisation

Cecília G. da Rocha, Carlos T. Formoso and Patrícia Tzortzopoulos

Abstract: Product modularity is a concept that can contribute to the improvement of product quality and production efficiency in house-building. However, there is a lack of consensus in the literature on the concepts that define product modularity. Furthermore, little attention has been given to the differences between building construction and manufacturing, for which product modularity was originally developed. This research aims to address that gap by adapting the conceptualization of product modularity so that it can effectively be used in the house-building industry. The methodological approach adopted in this study was Design Science Research, and two empirical studies were carried out on construction companies based in Brazil and in the U.K. Those studies are used to illustrate the applicability and utility of the proposed concepts and tools. Research findings indicate that the adoption of product modularity concepts results in benefits to both traditional construction technologies and pre-fabricated building systems.

Reprinted from *Sustainability*. Cite as: Rocha, C.; Formoso, C.; Tzortzopoulos, P. Adopting Product Modularity in House Building to Support Mass Customisation. *Sustainability* **2015**, *7*, 4919-4937.

1. Introduction

The entral idea of the mass customisation approach is to deliver products that have some degree of customisation while striving to meet the standards of efficiency, cost, and quality of mass production [1]. Therefore, it seeks to simultaneously achieve two seemingly conflicting goals, high product variety and high volume, by delivering products that fulfil the specific requirements of different customers through flexible processes and organizational structures [2]. In house building, mass customisation has been linked to environmental and social sustainability, as it can potentially eliminate waste resulting from changes carried out by occupants after moving in, and provide an opportunity for increasing their sense of identity and ownership.

Product modularity (or modular architecture) is one of the key elements of mass customisation strategies. It concerns the use of a limited set of modules to create several product variants [3]. The mixing and matching of modules in different combinations leads to high product variety, whereas high volume is achieved by using a limited number of modules across a large number of product variants [2].

The use of a set of interchangeable components sub-assemblies (modules) to build a wide range of dwellings has been pointed out as one of the necessary strategies to improve the performance of the house-building industry [4,5]. In this context, much of the design effort goes into designing modules that can be combined to produce a wide range of solutions [6]. This determines, to a large extent, the degree of customisation of dwellings and the level of efficiency that can be achieved in production [4,7,8]. Despite the potential benefits of product modularity, there are two main difficulties in the dissemination of this concept in the house-building sector.

The first difficulty relates to the lack of consensus on the conceptualization of product modularity. A large number of perspectives, languages and definitions are presented in the literature, from different fields of knowledge, such as engineering management, management science, and operations management [9]. According to Gershenson *et al.* [3], there is neither a clear measure of product modularity nor a widely adopted systematic method to help designers to increase the degree of product modularity. Such a diversity of definitions constitutes a major barrier to the advancement of scientific knowledge [9]. It hampers the creation of valid measures, which are necessary for empirical testing and theory building [9,10].

The second difficulty relates to the differences between the construction and manufacturing industries, as existing product modularity conceptualizations do not consider the peculiarities of construction, such as on-site production, temporary supply chains, and buildings as one-off-products [11]. In fact, previous studies investigated product modularity in construction, reinforcing the relevance of this topic in this sector. For instance, Voordijk *et al.* [5] explored the idea of aligning product, process and supply chain architecture, while Halman *et al.* [6] and Veentra *et al.* [12] investigated the development of product platform architectures for the house-building industry. Moreover, the concept of product modularity has a strong connection with the open building approach, which focuses on the introduction of different levels of decision making in the building process, and the possibility of decoupling building parts with different life cycles [13]. However, none of those studies have proposed a conceptual framework for product modularity for building products, addressing differences between manufacturing and construction.

This research aims to address that gap by adapting the conceptualization of product modularity so that it can effectively be used in the house-building industry. The research also proposes two operational tools, named the module combination matrix and module interactions matrix, which can support decision-making in the development of house-building products. Therefore, this study has a prescriptive character, as its main contribution focuses on understanding product modularity in this specific context to support the design of customised house-building projects. This investigation was part of broader research work whose main outcome was a conceptual framework for defining customisation strategies in house building [14].

2. Product Architecture and Modularity

Product architecture is a conceptual representation of the physical components used for making a product and the interactions between them, which affect the functioning of that product [15]. Any product has a product architecture regardless whether it was deliberately devised or not [16]. According to Salvador [9], the conceptualisation of product modularity does not point to a single concept, but it encompasses a number of different and interrelated concepts.

Ulrich [17] suggests that a product architecture is defined by three main decisions: (i) the arrangement of functional elements; (ii) the mapping from functional elements to physical components; and (iii) the specification of the interfaces among interacting physical components. The most common distinction made in the typology of product architecture is between a modular architecture and an integral architecture, although most products cannot be classified strictly as modular or integral [17]. There is rather a continuum, in which a product design can assume different architectures.

Functional elements define what a product does, whereas physical components are the chunks designed to fulfil these functions [17,18]. Clearly, there is a relationship between the functional elements and chunks of a product, termed as function-component allocation scheme [10]. Allocation schemes can be one-to-one, many-to-one, one-to-many, or many-to-many [17], depending on the design solution. In modular architecture, there is a one-to-one allocation scheme, in which each chunk (a sub-assembly of components) is physically independent and performs only one function [17]. Those chunks are termed modules [19] and can be mixed and matched in distinct combinations. By contrast, in integral architecture, there is a complex mapping (not one-to-one) from functional elements into physical components, and coupled interfaces [17]. This means that two or more physical components perform more than one function, or conversely two or more functions are performed by a single component.

The functional elements and components that form a product can be organized into hierarchical levels [9,19,20]. At the highest level, a function is provided by the whole product, whereas at the lowest level a very specific function might be provided by a single component [10]. Clearly, the level of interest when designing a product architecture is to define a set of sub-functions somewhere between these extremes [10], so that each function is close to client requirements, and the modules devised to fulfil these functions can be combined to meet distinct requirements.

Indeed, in mass customisation, product variants are often the result of different module combinations, which should be able to meet different client orders. Thus, it is important to devise modules that fulfil as many requirements as possible but that can also be used in a large number of product variants [10]. Pine II [2] described the main benefits of modular architecture as (i) economies of scale, which are achieved through the production of high volumes of standardised components rather than standardised products; (ii) economies of scope, which are achieved by using the modules over and over again in several product variants; and (iii) customisation being provided by using a limited number of modules to provide several product variants.

Platform is a concept closely related to modular architecture. A platform can be described as a basic common module that is used in several variants of a product family [21]. In other words, it is a specific module that encompass a larger chunk of physical parts than the other modules and that is used in every product variant that can be generated in a particular system [9].

A one-to-one allocation scheme is a necessary (although not a sufficient) condition for modular architecture [10]. In addition, the specification of interfaces needs to be appropriately established for modules to be interchangeably used across combinations [17]. The degree to which the interfaces among interacting modules follow a standard defines the compatibility among modules [22]. In other words, if two modules have the same interface standard it means that they can be combined. Using only one interface standard enables more modules combinations to be created than using several interface standards. Indeed, Salvador [9] points out that the highest level of combinability is achieved when all possible combinations of all available modules can be made. This can only occur when all modules have only one interface standard, as in sectional architecture [17]. However, such level of combinability might not be possible or desirable, and often there is more than one interface standard in a product architecture.

The number of different interface standards is a key issue when designing a product for mass customisation, as it determines the possible combinations between modules and, consequently, the product variants that can be created. The terms decoupled [17] or loosely coupled [12,23] are often used to describe modules that have the same interface standard and thus can be interchangeably used across different combinations. The terms coupled or tightly coupled refer to modules that cannot be easily combined either because of ambiguous allocation schemes or different interface standards [17].

The definition of interfaces between modules depends on the type of interaction that should exist between them. Pimmler and Eppinger [24] proposed a set of categories for interactions: (i) spatial interactions (*i.e.*, need for adjacency or orientation between two or more modules), energy interactions (*i.e.*, need for energy transfer between two modules), information interactions (*i.e.*, need for information exchange between two modules) and material interactions (*i.e.*, need for material exchange between two modules). Tsai and Wang [25] proposed a similar taxonomy of interaction types, adding another category, named geometric interactions, which refers to the fitting of two modules in relation to geometric parameters such as shape, size, and tolerance.

From a broader perspective, several decisions related to product design and production system design are required in the product development process. Consequently, other factors have a strong influence on whether products are delivered as a set of modules, which can be classified into three different domains: product, process, and supply chain [5,10]. For instance, Pahl *et al.* [26] make a distinction between function modules, *i.e.*, portions of the product implementing technical functions, and production modules, *i.e.*, portions of the product defined based on production considerations alone. Therefore, the product architecture is not fully defined by the design; it is also dependent on how a product is delivered.

3. Product Modularity in Construction

Voordijk *et al.* [5] pointed out the importance of analysing modularity in the construction industry from three perspectives: product, process and supply chain. These perspectives can be used to discuss the differences between construction and manufacturing that affect the adoption of product modularity in the delivery of buildings.

The main difference between manufacturing and construction from the product perspective relates to the nature of the parts that form the product itself. Most manufacturing products are simply divided into components, which have a central role in the definition of the product architecture. In this context, components are needed to perform specific functions that contribute to the function of the product at the highest level. For example, to enable a computer to perform its functions at the higher level, several sub-assemblies (e.g., monitor, keyboard, central processing unit, mouse) are needed. In contrast, the focus on components only is not appropriate for a building because buildings need to be considered as a mix of components and spatial voids. In fact, the main function of buildings, at the highest level, is not provided by components, but by the provision of spatial voids for people.

From the process perspective, the notion of modules as mechanically stable sub-assemblies, often involving a "glue-component" that keeps the parts attached [9], is another underlying notion of product modularity that is not completely applicable to buildings. In manufacturing, suppliers are able to deliver complex modules that are simply assembled by the main manufacturer [27], making this process simple. However, in most construction projects it is difficult to use this type of modules. Although mechanically stable modules can be applied in some highly industrialised projects, or in the production of some building subsystems in regular projects, much of the work in house-building is usually performed on site, using traditional technologies. This is strongly related to the unique character and immobility of buildings [11]. Therefore, much of the work still has to be performed on site, such as at the excavation and foundation stages. In addition, the finishing stages are often postponed and completed on site just prior to project delivery, in order to avoid damages in the finished product, which may be caused by the production activities that are taking place inside the building. As suggested by Voordijk *et al.* [5], in traditional building techniques, process architecture is dispersed in time but tight in space.

Finally, the unique and temporary nature of construction projects is another barrier to the design and production of modules as stable sub-assemblies. Construction project supply chains are essentially temporary multi-organizations that are set up with the specific purpose of delivering a construction project. Such supply chains arise at the start of the project, develop, and finally disband as the end of it [28]. In contrast with manufacturing, there is often limited incentive for suppliers to design and produce modules for a large number of projects. This is also related to the fact that building projects are one-of-a-kind, which means that the modules developed for a project are unlikely to be suitable for other projects [4].

Table 1 summarizes the differences between construction and manufacturing to be considered in product modularity.

Table 1. Differences between construction and manufacturing relevant for product modularity.

Perspectives	Manufacturing	Construction
Product	Components have a central role in the definition of the product architecture	Buildings combine components and spatial voids, which perform the most important product functions
Process	Suppliers deliver complex modules that are simply assembled by the main manufacturer	Much work is usually performed on site, using traditional technologies
Supply chain	The supply chain gets involved in the design and production of modules for a large number of products	Temporary supply chains usually have limited incentives to produce the same module for a large number of projects

4. Research Method

The methodological approach adopted for the development of the framework was design science research, which is a form of scientific knowledge production that involves the development of innovative constructions, intended to solve problems faced in the real world, and simultaneously makes a prescriptive scientific contribution [29]. An important outcome of this type of research is an artifact that solves a domain problem, also known as solution concept, which must be assessed against criteria of value or utility [30]. Even though this approach was developed originally in the area of Information Systems, a number of authors, such as Voordijk [31] and AlSehaime *et al.* [32], suggest that design science research should be used to produce appropriate multidisciplinary solution concepts for solving complex and relevant problems in the Built Environment.

In this research, the proposed artifact is a set of constructs that provide an understanding of product modularity in house-building. This set of concepts emerged from two case studies carried out in different house-building companies, through several cycles of development, testing and refinement, until a suitable version was produced. Both companies were chosen because they delivered customised housing units, and also due to their willingness to participate in this investigation. One of them used a highly industrialised building system (Company B), and the other adopted fairly traditional construction technologies (Company A).

Company A is a Brazilian contractor that develops and builds apartment buildings for high-end clients. Most of the construction methods adopted by this company involve activities that are carried out on site, such as cast in place concrete structure, masonry walls, and plastering. Data from one specific project, a high-rise residential project, named Project J, was analysed in this investigation.

Company B is a British firm that builds residential schemes for social housing using a prefabricated system. The system consists of a set of timber-framed pods produced in a factory. Each pod includes a set of rooms that is fully fitted, with windows and doors in place. After the pods are delivered to the site, the external finishing and roof are assembled, completing the units.

The aim of carrying out multiple case studies was to assess the utility of the set of concepts in two very different contexts. In each case, an assessment of the building design process in terms of product modularity was developed, and improvement opportunities were identified. Such assessment is in line with the design science research approach, which requires the appraisal of the practical contribution of the solution [33,34].

Multiple sources of evidence and a causal chain of evidence were used to create valid constructs, and a case study protocol was adopted to ensure the reliability of data collection, as suggested by Yin [35]. However, differently from descriptive research strategies from social sciences (like case studies), the focus of this investigation was the development of an artefact, the conceptualization of which is presented in Section 5. Thus, the role of the case studies was to test the utility of the set of concepts that have been proposed, as opposed to pattern matching, which should be used if the research strategy was multiple case studies.

Table 2 describes the sources of evidence used in this research study. Based on the data collection, the following information was produced for each company: (a) description of the business model, competitive criteria, and the profile of its main clients; (b) an overall map of the design and construction process; (c) identification of existing customization practices, especially those related to product modularity; and (d) description of the main products delivered and customisation options.

Table 2. Sources of evidence and focus of data collection in Case Studies A and B.

Case Study A	Case Study B
- 13 semi-structured interviews with staff involved in the design and construction processes - Analysis of design documents, including strategic plans, architectural drawings and commercial leaflets - Participant observation in two meetings involving top and middle management - Direct observation of the construction process on site	- 8 semi-structured interviews with staff involved in the design and construction processes - Analysis of design documents, including architectural drawings and commercial leaflets - Direct observation of the construction process in the manufacturing plant - One meeting with top and middle management to discuss the outcomes of the research

5. Proposed Conceptualisation of Product Modularity in House-Building

The proposed conceptualisation includes three elements: (i) product architecture perspectives; (ii) interfaces among interacting modules; and (iii) operational tools to support decision-making. Each part is described as follows.

Regarding the product architecture perspectives, there are two fundamental types of functional elements (primary and secondary functions) and physical components (spatial voids and solid mass) in buildings. The terms “spatial voids” and “solid mass” are borrowed from Ching [36] who uses them to explain architectural forms. Primary functions (e.g., reading, sleeping, eating, and working) are performed by people in spatial voids. Secondary functions (e.g., load support, enclosure, roof, and finishings) are performed by the solid mass of a building. Spatial voids and solid mass are the fundamental parts of any building. Therefore, the solid mass components of a building have two main roles: (i) to create the spatial voids; and (ii) to enable the primary functions to be comfortably carried out by people.

Some building components fulfil only one of these roles, whereas others fulfil both of them. For example, components forming the structural system essentially create the spatial voids. Building services, such as water, electricity, and heating, do not create the spatial voids but support primary functions. Other building parts such as the enclosure system fulfil both roles as they create the spatial voids and also protect these spatial voids from the weather.

Although spatial voids and solid mass are complementary, the product architecture could be conceived in terms of solid mass (components perspective) or spatial voids (spatial perspective). In fact, each of these perspectives embodies a different type of allocation scheme. From a spatial perspective, the focus is on the primary functions performed by people in the spatial voids (Figure 1). However, these functions still need to be allocated to some extent into solid mass, as these are needed to create the spatial voids. In a components perspective, the focus is on the allocation of secondary functions into solid mass (Figure 1).

		Functional elements	
		Primary functions	Secondary functions
Physical elements	Spatial voids	**Spatial Perspective**	
	Solid mass (Components)		**Components Perspective**

Figure 1. Two perspectives on the product architecture of buildings.

The components perspective is implicit in the product architecture design of most manufactured products, as product components, rather than people, perform most functions. For example, Stone *et al.* [37] proposed a method to identify modules in manufactured products, in which classes of functions are essentially performed by components. Clearly, the spatial perspective can be adopted for other products that involve spatial voids such as ships and airplanes, and even for cars. However, the network of spatial voids in some of those products is likely to be simpler than the one found in buildings, as the size and number of spatial voids tends to be much smaller.

In terms of interactions among modules, the categories proposed by Pimmler and Eppinger [24] and Tsai and Wang [25], described in Section 2, are applicable to buildings. However, it seems that spatial interactions and geometric interactions play a key role. This is because the spatial voids and solid mass forming a building need to be geometrically and spatially coordinated to enable the primary functions to be appropriately carried out. A geometrical interaction involves the physical fitting between two modules. For example, the sink piping and the drain should have a particular geometry so that they can be connected. A spatial interaction involves the spatial coordination between two modules. For example, a sink should be positioned at a particular height in relation to the floor.

The other types of interactions discussed by Pimmler and Eppinger [24] and Tsai and Wang [25] apply only to some building components. For example, there are load exchanges (energy transfer) among modules that form the structural system, and water exchanges (material flow) among modules that form the water and sewage system. However, most spatial and geometric interactions also apply to them and in fact are needed to support the other interactions. In the sink example, the geometric interaction between the sink piping and the drain need to be appropriately established for the water to flow. Similarly, the beams and columns need to be geometrically and spatially coordinated for the load to be adequately transferred.

Two fairly simple tools have been devised to analyse product architecture in house buildings: (a) the module combinations matrix; (b) the module interactions matrix.

The module combinations matrix clearly describes the modules used across diverse product variants in a visual format, as shown in Table 3. The matrix also incorporates a metric, called the module use index (MUI) to assess the degree of module usage, which is a proxy of commonalities across product variants: a high MUI means that many modules are common to several product variants. The higher the MUI, the larger the economy of scope, as it represents greater module usage.

This, in turn, leads to a simplified production process. The MUI can be used to assist designers to evaluate and compare the economies of scope for different design solutions.

Table 3. Module Combinations Matrix of a hypothetical product architecture

	Product Variant A	Product Variant B	Product Variant C	Number of Uses
Module 1	2		1	3
Module 2		1	2	3
Module 3	1			1
Module 4		1	1	2
			MUI	2.25 (9/4)

The MUI can also be used to assess whether a delayed product differentiation could be adopted for a particular product design. Postponing product differentiation until the last possible moment enables organisations to operate with efficiency and to quickly meet clients' orders with minimum amount of stock [38]. With such an approach, product variants are designed using common platforms and modules and are assembled only when a client order is received [39]. Commonality, in turn, dictates to a large extent the success of a delayed product differentiation approach [40]. If a small number of modules is used across several product variants, the modules can potentially be produced based on forecast since they will probably be used in any upcoming order. This is particularly important if there is an off-site manufacturing system that provides modules (e.g., pods, components) to different construction sites.

The module interactions matrix displays the interfaces that exist between modules in each of the product variants (e.g., A, B, C in Table 4). It indicates the number of interactions between modules in a product family. This matrix can be used to identify the need for improving interfaces between modules. Both matrices indicate whether any of the modules can be considered as a platform. A platform module is found in all product variants, and must be able to interact with any other modules.

Table 4. Module interactions matrix of a hypothetical product architecture.

	Module 1	Module 2	Module 3	Module 4
Module 1	A			
Module 2	C			
Module 3	A			
Module 4		B, C		

The application of these concepts and tools in the case studies is described below.

6. Results

6.1. Case Study A

The apartment buildings developed by Company A can be organized into four product families. Each product family has a particular householder profile, an average apartment size, and a degree of customisation. In Project J, this company intended to offer buyers the possibility of choosing the apartment internal layout as well as the floor tiles for the social areas. For each of these attributes (floor tiles and layout), Company A planned to offer a limited set of options for clients to select from.

The choice of floor tiles is a low hierarchical level decision in the product architecture, since the function (floor finishing) is provided by only one component type (floor tile). This type of customisation does not provide a major problem in terms of product architecture as difficulties most often arise at higher levels, where more than one component type provide a function.

By contrast, the choice of floor plan is at a high hierarchical level in the product architecture design as the function (layout) is provided by several components. Product variants are created by altering the primary functions of some spatial voids (and the solid mass associated to it) while keeping the rest of the apartment unchanged. In Project J, there are three product variants: 3-bedroom apartment with large living room (product variant X), 3-bedroom apartment with TV lounge (product variant Y), and 4-bedroom apartment (product variant Z). They are provided by changing the primary functions of four rooms (A, B, C, and D)—see Table 4 and Figure 2.

The rest of the apartment configures an additional module. It can be regarded as a platform, as it encompasses a considerable part of the product and is used across the three product variants. The MUI is 1.50, because three modules (5, 8, and 10) are used twice and the platform is used three times across the module combinations (Table 5). As shown in Table 6, all modules interact (have interfaces) with the platform as they are located inside it. The modules used in rooms A and B also interact due to their physical adjacency, similarly to the modules used in rooms C and D (Figure 2).

Table 5. Module Combinations Matrix—Case Study A.

		Product Variant X	Product Variant Y	Product Variant Z	Number of Uses
	Module 1	1	1	1	3
Room A	**Module 2**	1			1
	Module 3		1		1
	Module 4			1	1
Room B	**Module 5**	1		1	2
	Module 6		1		1
Room C	**Module 7**	1			1
	Module 8		1	1	2
Room D	**Module 9**	1			1
	Module 10		1	1	2
				MUI	**1. 50** (15/10)

Table 6. Module Interactions Matrix—Case Study A.

	Module 1	Module 2	Module 3	Module 4	Module 5	Module 6	Module 7	Module 8	Module 9	Module 10
Module 1										
Module 2	X									
Module 3	Y									
Module 4	Z									
Module 5	X, Z	X		Z						
Module 6	Y		Y							
Module 7	X									
Module 8	Y, Z									
Module 9	X						X			
Module 10	Y, Z							Y, Z		

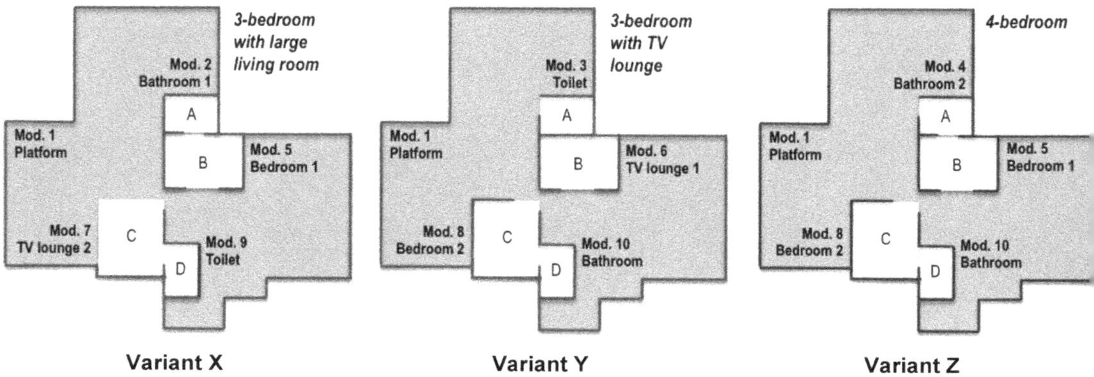

Figure 2. Diagrammatic footprint of product variants X, Y and Z.

A major problem identified in Project J is the ambiguous allocation scheme. The internal walls that enclose the modules are the same walls enclosing the rooms that form the platform, which means that some components pertain to more than one module. These components can only be built once the modules have been defined, which may happen only after the client has selected a product variant (an apartment floor plan). This ambiguous allocation scheme also creates coupled interfaces and, thus, physical changes are needed to fit the different modules into the platform.

Another problem is that the platform was not designed to enable the modules in rooms A, B, C and D to be used interchangeably (Figure 2). For instance, in order to build module 2 in room A, the platform should have a particular layout for the building service systems that is different from the layout needed for module 3. Likewise, modules 4 and 5 have different electrical layouts (*i.e.*, lights, switches, and plugs positioning) and the platform needs to be modified depending on the modules used. Designing a one-to-one allocation scheme is the first step to solve these problems. In addition, the platform and modules should be redesigned to ensure that they could be interchangeably used. These changes would ease the adoption of a delayed product differentiation.

6.2. Case Study B

There were six customisable attributes in the pre-fabricated building system in case study B: (i) kitchen fit outs; (ii) bathroom fit outs; (iii) windows and doors; (iv) facades; (v) roofing; and (vi) dwelling types. The fit-out options entail different specifications for bathroom and kitchen finishings and thus involve a decision at a lower hierarchical level in the product architecture. The scope of customisation embedded in windows, doors, facades and roofing was not clearly defined by Company B. For example, for windows and doors it was not clear if changes entailed (i) the amount of windows; (ii) size of windows; (iii) location of windows in the facade, or all of the above. This hindered the analysis of the implications of some attributes in the product architecture. Nonetheless, "housing types" was the customisable attribute with the most important implications in product architecture. This is because it involves a function at a high hierarchical level, which is provided by a sub-assembly of components (*i.e.*, pod). As shown in Figure 3, ten pods organised into three sizes are used to provide five product variants (dwelling types).

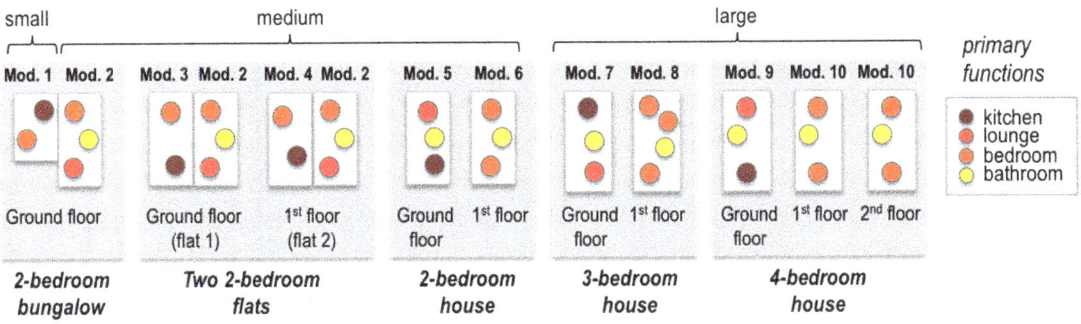

Figure 3. Diagrammatic footprint of the pods.

Each pod can be viewed as a module. The pods are combined one on the top of the other or side-by-side to create different product variants (Figure 3): a 2-bedroom bungalow (product variant A), two 2-bedroom flats (product variant B), a 2-bedroom house (product variant C), a 3-bedroom house (product variant D), and a 4-bedroom house (product variant E). Therefore, each pod has a set of primary functions, suggesting that a spatial perspective was adopted in devising this product architecture. The MUI has a low result (1.30) for an industrialised building system; module 2 is used three times and module 10 twice, whereas the other modules are used only once in any combination (see Table 7).

All modules interact with only one module, except for modules 2, 3, 4 and 10 (see Table 8). Modules do not require physical changes to be used across the different product variants (see Table 7), indicating that their interfaces are decoupled. Also, this product architecture has a one-to-one allocation scheme between the primary functions and the components that support them. Consequently, no building components pertain to two or more modules. This allocation scheme together with the decoupled interfaces enable each module to be completely built regardless of the other modules that it will be combined with.

Table 7. Module combinations matrix—case study B.

	Product Variant A	Product Variant B	Product Variant C	Product Variant D	Product Variant E	Number of Uses
Module 1	1					1
Module 2	1	2				3
Module 3		1				1
Module 4		1				1
Module 5			1			1
Module 6			1			1
Module 7				1		1
Module 8				1		1
Module 9					1	1
Module 10					2	2
					MUI	**1. 30** (13/10)

Table 8. Module Interactions Matrix—Case Study B.

	Module 1	Module 2	Module 3	Module 4	Module 5	Module 6	Module 7	Module 8	Module 9	Module 10
Module 1										
Module 2	A	B								
Module 3		B								
Module 4		B	B							
Module 5										
Module 6					C					
Module 7										
Module 8							D			
Module 9										
Module 10									E	E

Five different modules are needed to create the two 2-bedroom dwellings, namely, the 2-bedroom-house and the two 2-bedroom-flats (see Figure 3). However, both dwellings have the same functional requirements in terms of spatial voids as they lodge the same number of householders. Based on that, different combinations of modules were simulated seeking to identify opportunities to increase the MUI. An important conclusion was that modules 5 and 6 could also be used to create a two 2-bedroom-flat (see Figure 3). Furthermore, module 6 could replace module 2 in the 2-bedroom-bungalow. In this way, modules 2, 3, and 4 could be eliminated, increasing MUI from 1.30 to 1.86 without affecting the product variants offered, as illustrated in Table 9. This new combination is also more suitable than the existing one as it enables the adoption of a delayed product differentiation, since the five product variants can be generated using fewer modules. This shows that even a highly modular architecture (*i.e.*, one-to-one allocation scheme and decoupled interfaces) such as this can benefit from a careful analysis of module combinations.

Table 9. Module combinations matrix (based on suggested changes)—case study B.

	Product Variant A	Product Variant B	Product Variant C	Product Variant D	Product Variant E	Number of Uses
Module 1	1					1
Module 5		2	1			3
Module 6	1	2	1			4
Module 7				1		1
Module 8				1		1
Module 9					1	1
Module 10					2	2
					MUI	**1. 86** (13/7)

The discussions with Company B representatives made them aware that the existing product architecture had ten modules instead of three, as originally assumed. The representatives had assumed the existence of three modules only because the pod had three standard sizes (small, medium and large as shown in Figure 3). Based on this new understanding, Company B realized that it was necessary to investigate alternatives to increase the commonalties across the combinations in order to reduce the overall number of modules. These plans were driven by the analysis of the product architecture and also by the improvements in the combination of modules proposed in this research. Even though some refinements in the design of modules 5 and 6 were necessary to enable their reuse across the product variants, the analysis and proposed changes were perceived as very positive by the company. This illustrates how the proposed conceptualisation was helpful in pointing out how the existing product architecture could be enhanced.

6.3. Discussion

A summary of the results for each case study is presented in Table 9. Case study A presented an ambiguous allocation scheme and coupled interfaces, issues that could be solved through the creation of a one-to-one allocation scheme and assigning each component to only one module (Table 9). Case study B had a highly modular product architecture, but the MUI was low, indicating an opportunity to improve the modules combination (Table 10).

Both components and spatial product architecture perspectives were considered in each case study. The components perspective is not novel, as it is described in most product modularity studies in manufacturing. This perspective has been previously adopted also in house-building design. For example, the product architecture described by Hofman *et al.* [4] adopts a component perspective since it involves three module types (columns, exterior walls, and floors) that fulfil secondary functions, and which are mixed and matched to created different building types. Conversely, the spatial perspective has not been widely investigated in house-building. In that perspective, each module involves a wide assortment of interconnected components. Consider for instance the different components needed (e.g., floor, ceiling, walls, fixtures, finishing) to create a module that supports the primary functions associated with a bedroom (e.g., to read, to sleep, to watch TV). In a spatial perspective, modules should involve sizeable portions of the product, resulting in fewer modules and interfaces. These fewer modules and interfaces are beneficial from a production viewpoint, due to the reduction in the number of modules that need to be produced. However, devising a modular architecture defined by a one-to-one allocation scheme and decoupled interfaces can be challenging, because each spatial void requires several components.

Table 10. Summary of the case study results.

	Case Study A	Case Study B
Product architecture perspectives	- Spatial perspective in product architecture design - Ambiguous allocation scheme between functional elements and physical components - Some building components pertains to more than one module	- Spatial perspective in product architecture design - One-to-one allocation scheme between functional elements and physical components - Each building component pertains to only one module
Interfaces among interacting modules	- Coupled interfaces: changes are required in the modules depending on the product variant they are used in	Decoupled interfaces: modules can be used across the product variants without changes
Operational tools to support decision-making	- Use of a platform (module 1) and nine other modules to create three product variants - All modules interact with two or more modules - Module Use Index (MUI) = 1.50 (original)	- Use of ten modules (no platform) to create five product variants - Most modules interact with only one other module (except for modules 2, 3, 4 and 10) - Module Use Index (MUI) = 1. 30 (original) - Module Use Index (MUI) = 1. 86 (for suggested changes in modules combinations)

In addition, each perspective involves distinct relationships between functional elements and physical components. These differences need to be recognized, as failure to do so could lead to product architectures that provide limited support to mass customisation. For example, in case study A, standard sizes were used for modules so that they could interchangeably fit in the rooms. Thus, attention was given to the dimensions of the spatial voids but not to the components needed to create these voids, as indicated by the ambiguous allocation scheme in which components pertain to two or more modules, illustrated in Table 10. The lack of understanding of the relationship between the elements involved is the root cause of this problem; spatial voids require components to exist, and these need to be associated to only one module. This highlights the importance of understanding the differences between those two perspectives as a first step to overcome potential product development problems.

The application of the conceptualization proposed in the case studies suggest that it can be used to support the design of customised house-building projects, both for the identification of problems and to support the design of solutions for better house-building customisation strategies. Further empirical evidence of the usefulness of the concepts and tools proposed in this research was also provided by the fact that company B planned actions to improve its product architecture based on the analysis provided.

7. Conclusions

The adoption of product modularity in house-building design can contribute to the implementation of the mass customisation approach in this industry, despite the differences between manufacturing and construction. The main theoretical contribution of this investigation is the adoption of two different perspectives on product architecture that should be considered in the design of house-building projects, namely the component perspective and the special perspective. While the first perspective is widely adopted in the manufacturing industry, the spatial perspective is necessary due to the fact that the primary functions of buildings are performed by people in spatial voids, rather than by solid mass.

This study also emphasized that spatial and geometric interactions play a key role in product modularity in house-building, because the spatial voids and solid mass forming a building need to be geometrically and spatially coordinated to enable the primary functions to be appropriately carried out. Moreover, spatial and geometric interactions are needed to support other interactions, such as load exchanges or water.

As a practical contribution, two tools for assessing product architecture of residential buildings have been devised, the module combinations matrix and the module interactions matrix. Based on the first one, a metric, called the module use index, can be used to evaluate and compare the economies of scope for different design solutions.

Case study A indicated that product modularity concepts can be beneficial even for companies that use mostly traditional construction technologies, often performed on site. The problems identified in that study are related to inappropriate design decisions concerning the modules and module interfaces and are thus unrelated to the construction method selected.

Case study B illustrated that the use of modules that are stable sub-assemblies, which facilitate the design of a one-to-one allocation scheme and decoupled interfaces, is not a sufficient condition to get full benefits from product modularity when a mass customisation approach is adopted. That study highlighted the importance of analysing module combinations when designing a product for mass customisation. This reinforces the usefulness of the module combination matrix and module interactions matrix, which were proposed in this investigation, as they can be used to assess different product families.

Regarding future studies, at least three topics can be explored. First, this paper has mainly focused on spatial and geometrical interactions among modules and it is therefore still necessary to explore other types of interaction and their incidence in buildings. Second, the characteristics of modules and products variants should be incorporated into upcoming reuse indexes. This is important to better assess the benefits yielded by module reuse (e.g., the reuse of large modules such as the platform in case study A are likely to yield more benefits than the reuse of small modules). Third, further studies should investigate the usefulness of the proposed concepts and tools in the development of new house-building projects.

Acknowledgments

The authors would like to thank the National Council of Technological and Scientific Development (CNPq) and Coordenação de Aperfeiçoamento de Pessoal de Nível Superior (CAPES) for funding this research study. They would also like to thank Carl Abbot and Will Swan from the University of Salford and Mike Kagioglou from the University of Huddersfield for supporting the development of this research.

Author Contributions

All three authors were equally involved in the writing of the paper. Cecília da Rocha and Carlos Formoso were involved in the research design, and Cecília Rocha carried out data collection and processing. All authors have read and approved the final manuscript.

Conflicts of Interest

The authors declare no conflict of interest.

References

1. MacCarthy, B.; Brabazon, P. In the business of mass customisation. *IEEE Manuf. Eng.* **2003**, *82*, 30–33.
2. Pine, B.J., II. Mass Customizing Products and Services. *Strategy Leadersh.* **1993**, *21*, 6–55.
3. Gershenson, J.K.; Prasad, G.J.; Zhang, Y. Product modularity: Definitions and benefits. *J. Eng. Des.* **2003**, *14*, 295–313.
4. Hofman, E.; Voordijk, H.; Halman, J. Matching supply networks to a modular product architecture in the house-building industry. *Build. Res. Inf.* **2009**, *37*, 31–42.
5. Voordijk, H.; Meijboom, B.; de Haan, J. Modularity in supply chains: A multiple case study in the construction industry. *Int. J. Oper. Prod. Manag.* **2006**, *26*, 600–618.
6. Halman, J.I.M.; Voordijk, J.T.; Reymen, I.M.M.J. Modular approaches in Dutch house building: An exploratory survey. *Hous. Stud.* **2008**, *23*, 781–799.
7. Barlow, J. From Craft Production to Mass Customisation. Innovation Requirements for the UK Housebuilding Industry. *Hous. Stud.* **1999**, *14*, 23–42.
8. Noguchi, M. The effect of the quality-oriented production approach on the delivery of prefabricated homes in Japan. *J. Hous. Built Environ.* **2003**, *18*, 353–364.
9. Salvador, F. Toward a Product System Modularity Construct: Literature Review and Reconceptualization. *IEEE Trans. Eng. Manag.* **2007**, *54*, 219–240.
10. Fixson, S.K. Product architecture assessment: A tool to link product, process, and supply chain design decisions. *J. Oper. Manag.* **2005**, *23*, 345–369.
11. Koskela, L. Is structural change the primary solution to the problems of construction? *Build. Res. Inf.* **2003**, *31*, 85–96.
12. Veenstra, V.S.; Halman, J.I.M.; Voordijk, J.T. A methodology for developing product platforms in the specific setting of the housebuilding industry. *Res. Eng. Des.* **2006**, *17*, 157–173.
13. Kendall, S.H.; Teicher, J. *Residential Open Building*; Taylor and Francis: Oxon, UK, 2010; p. 320.

14. Rocha, C.G. A Conceptual Framework for Defining Customisation Strategies in the House-Building Sector. Ph.D. Thesis, Federal University of Rio Grande do Sul, Porto Alegre, Brasil, 2011.
15. Wang, H.; Kimble, C. Low-cost strategy through product architecture: Lessons from China. *J. Bus. Strategy* **2010**, *31*, 12–20.
16. Martin, M.V.; Ishii, K. Design for variety: Developing standardized and modularized product platform architectures. *Res. Eng. Design* **2002**, *13*, 213–235.
17. Ulrich, K. The role of product architecture in the manufacturing firm. *Res. Policy* **1995**, *24*, 419–440.
18. Dahmus, J.B.; Gonzalez-Zugasti, J.B.; Otto, K.N. Modular product architecture. *Design Stud.* **2001**, *22*, 409–424.
19. Muffatto, M. Platform strategies in international new product development. *Int. J. Oper. Prod. Manag.* **1999**, *19*, 449–459.
20. Salvador, F.; Forza, C.; Rungtusanatham, M. Modularity, product variety, production volume, and component sourcing: Theorizing beyond generic prescriptions. *J. Oper. Manag.* **2002**, *20*, 549–575.
21. Blecker, T.; Abdelkaf, N. Complexity and variety in mass customization systems: Analysis and recommendations. *Manag. Decis.* **2006**, *44*, 908–929.
22. Mikkola, J.H. Capturing the Degree of Modularity Embedded in Product Architectures. *J. Prod. Innov. Manag.* **2006**, *23*, 128–146.
23. Sanchez, R.; Mahoney, J.T. Modularity, flexibility, and knowledge management in product and organization design. *Strateg. Manag. J.* **1996**, *17*, 63–76.
24. Pimmler, T.U.; Eppinger, S.D. Integration analysis of product de-compositions. In Proceedings of the ASME Conference on Design Theory and Methodology, Minneapolis, MN, USA, 11–14 September 1994; pp. 343–351.
25. Tsai, Y.; Wang, K.; Lo, S. A study of modularity operation of systems based on maintenance considerations. *J. Eng. Des.* **2003**, *14*, 41–56.
26. Pahl, G.; Beitz, W.; Feldhusen, J.; Grote, K.H. *Engineering Design: A Systematic Approach*, 3rd ed.; Springer: London, UK, 2007; p. 617.
27. Alford, D.; Sackett, P.; Nelder, G. Mass customisation–An automotive perspective. *Int. J. Prod. Econ.* **2000**, *65*, 99–110.
28. Cherns, A.; Bryant, D. Studying the clients role in construction management. *Constr. Manag. Econ.* **1983**, *177*, 177–184.
29. Lukka, K. The constructive research approach. In *Case Study Research in Logistics*; Turku School of Economics and Business Administration, Series B1; Ojala, L., Hilmola, O-P., Eds.; University of Turku: Turku, Finland, 2003; pp. 83–101.
30. March, S.T.; Smith, G.F. Design and natural science research on information technology. *Decis. Support Syst.* **1995**, *15*, 251–266.
31. Voordijk, H. Construction management and economics: The epistemology of a multidisciplinary design science. *Constr. Manag. Econ.* **2009**, *27*, 713–720.

32. AlSehaimi, A.; Koskela, L.; Tzortzopoulos, P. Need for Alternative Research Approaches in Construction Management: Case of Delay Studies. *J. Manag. Eng.* **2013**, *29*, 407–413.
33. Kasanen, E.; Kari, L.; Siitonen, A. The constructive approach in management accounting research. *J. Manag. Account. Res.* **1993**, *5*, 243–264.
34. Van Aken, J.E. Management research based on the paradigm of the design sciences: The quest for field-tested and grounded technological rules. *J. Manag. Stud.* **2004**, *41*, 219–246.
35. Yin, R. Case Study Research: Design and Methods, 4th ed.; Sage: Los Angeles, CA, USA, 2013.
36. Ching, F. *Architecture: Form, Space, and Order*, 3rd ed.; Wiley: New York, NY, USA, 2007; p. 448.
37. Stone, R.B.; Wood, K.L.; Crawford, R.H. A heuristic method for identifying modules for product architectures. *Des. Stud.* **2001**, *21*, 5–31.
38. Feitzinger, E.; Lee, H.L. Mass customisation at Hewlett-Packard: The power of postponement. *Harvard Bus. Rev.* **1997**, *75*, 116–123.
39. Christopher, M. The Agile Supply Chain: Competing in Volatile Markets. *Ind. Mark. Manag.* **2001**, *29*, 37–44.
40. Piller, F.; Kumar, A. For each, their own: The strategic imperative of mass customization. *Ind. Eng.* **2006**, *38*, 40–45.

A Construction Management Framework for Mass Customisation in Traditional Construction

María Dolores Andújar-Montoya, Virgilio Gilart-Iglesias, Andrés Montoyo and Diego Marcos-Jorquera

Abstract: A Mass Customisation model is discussed as a competitive positioning strategy in the marketplace adding value to the customer's end-use. It includes the user as part of the construction process responding to the customer's demands and wishes. To the present day, almost all proposals for Mass Customisation have been focused on the design phase and single family houses. The reality is that the processes carried out in the work execution are so inefficient that the costs of the Mass Customisation models are assumed by the customer and they do not offer solutions that support the change management. Furthermore, this inefficiency often makes Mass Customisation unfeasible in terms of deadlines and site management. Therefore, the present proposal focuses on achieving the paradigm of Mass Customisation in the traditional residential construction complementary to the existing proposals in the design phase. All this through the proposal of a framework for the integral management in the work execution, which will address change management introduced by the users offering an efficient and productive model that reduces costs in the process. This model will focus on the synergy between different strategies, techniques and technologies currently used in the construction management (such as Lean Construction or Six Sigma), together with, other strategies and technologies that have proven to be valid solutions in other fields (such as Business Process Management, Service Oriented Architecture, *etc.*).

Reprinted from *Sustainability*. Cite as: Andújar-Montoya, M.; Gilart-Iglesias, V.; Montoyo, A.; Marcos-Jorquera, D. A Construction Management Framework for Mass Customisation in Traditional Construction. *Sustainability* **2015**, *7*, 5182-5210.

1. Introduction: Mass Customisation in Housing Market

Mass Customisation is emerging as an inevitable strategy to ensure competitiveness in a customer-oriented market [1], this means to satisfy more closely the individual wants and needs of the customers, by providing a wide variety of options and individual customisation at prices comparable to standard goods [2]. So, mass customization is the capability to integrate these varieties derived from the individual customer's needs and the efficiency of standard mass production [3,4]. Specifically, in the Spanish housing sector the need of adopting competitive strategies oriented to the prospective buyer is more pronounced due to the current economic slowdown, the sharp drop in activity and the oversupply in dwelling that difficulty the home sales. The strong adjustment of the sector, the high number of dwellings in stock, the largest social requests from the future users and the increased normative requirements, especially in terms of sustainability, safety and quality in buildings, address companies to the imminent need of being reinforced to compete in the market [5]. Facing this new development, enterprises need to be positioned in the market through a differentiation with competitors by implementing new ways to add value to their offers [6]. Through the offer of Mass Customisation in the property development, value-added services are created replying to the need of some potential customers who are changing the way they used to buy, integrating customer into the process. So a new agile and flexible approach is needed not only to satisfy the initial customers' requirements, but also to provide a greater ability to adapt to their needs as they evolve [7] with the same cost as in mass production [8].

However, the barriers to achieving this approach in the field of building and construction are high. At the same time that enterprises are forced to react to the growing individualization of demand, the increased pressure of competitors and the financial conditions of the market dictate the need of shorter construction time, reduction in total costs and higher production quality. Companies must adopt strategies that include both cost efficiency and a closer relationship to the customer's needs [9]. This need of efficiency in building costs is one of the principal barriers to Mass Customisation, as it leads to increased material costs, higher manufacturing costs, lower on time deliveries, increased level of needed inventory for every variant of Customisation and reduction in product quality [1]. These increases in costs cannot be absorbed by customers who usually expect individualized products at the same price as they would pay for mass-produced items [10].

Besides the increase in cost, Mass Customisation accompanies problems related to term, due to the fact that identical houses cannot be produced. Construction firms require more time to carry out the site works for not being totally familiarity with the plans, owing the variability of them from some property to another [11].

Furthermore, Ph.D. Noguchi describes in "ZEMCH Research Initiatives: Mass Customisation and Sustainability" [12] the additional obstacles related to customisation in property developments, such as lack of pilot housing, more complexity in communication, greater chance of material delays, and disadvantages in negotiations with subcontractors.

Another obstacle attributed to the housing customisation is the delay in the buyer decision making, a fact that can cause delays in the planning of works [13]. Also the risk of doubtful customers and changes in opinion derived from the array of choices.

In addition, this flexibility and variability increase the complexity of the management and the execution of the whole process given that the diversity of options may exceed the capacity of the organization to manage works efficiently.

Currently, construction companies do not have information systems to address efficiently the management of Mass Customisation in site. This obstacle is compounded by the inefficiency in productivity indexes that has been dragging the industry in the last years, in comparison with other sectors of the Spanish economy.

Consequently of the foregoing and considering the current situation of low profit margins, the main barrier to adopt Mass Customisation in housing is the lack of a structured system to help the site management efficiently, minimizing the extra costs, the deviations in schedules, the materials delays and the uncertainty in the planning. Therefore, the short-term challenge for the industry is to increase efficiency by reducing these barriers at the same time the communication and negotiations are improved bringing simplicity to the management, and providing added value to the product.

This requires the adoption of a new management model that makes feasible Mass Customisation in property developments, allowing the satisfaction of the customer requirements, thus accentuating the competitive advantage of firms.

At the present time there are proposals with large repercussion to reinforce company competitiveness, and then overcome the main barriers to adopt the Mass Customisation. They are mainly centred on solving the inefficiency in cost and time of the construction works, while they focus on meeting the customers' needs. To develop and emphasize the competitive advantage of a corporation some construction companies are starting to carry out relevant business management strategies to strength their position in the market. One of the most emerging approach to improve organizational effectiveness is the application of Lean Philosophy [14] to construction, which is consider as a part of a "cost leadership [15]" competitive strategy [16] that maximize customer value while minimizing waste. Also another upward trend to increase productivity is the implementation of Information and Communication Technology (ICT) as a competitive approach that is increasingly being used in the construction industry [17]. Several options of specific software for the works are available to increase productivity in the construction process (e.g., software for quality control, planning, project based accounting, *etc.*). Also software tools of enterprise performance systems (ERP) or customer relationship management (CRM), among others, are implemented by many companies.

Therefore to exploit the whole potential of each specific proposal it should be approached in a holistic solution [18], avoiding the technological waste with the "point solutions" for each stage of the construction process and looking for a well-suited solution to construction [19,20]. Accordingly, to get a holistic solution in the site execution, the main novelty lies in using for the first time, Business Process Management (BPM) in construction, in particular to provide an integrated management model by using BPM as a backbone solution. BPM is an evolving trend in management science whose effectiveness has been commonly demonstrated in other fields such as the industry sector, banking, retail, government, and health care. The importance of BPM is confirmed by the existence of specialized international journals (e.g., the Business Process Management Journal), conferences (e.g., the Business Process Management Conference) and courses institutionalized at several universities [21]. This is a powerful instrument to gain competitive advantage through a holistic process-oriented view [22] based on the ICT to automate or support processes and their lifecycle. This holistic view can involve a lot of management disciplines and techniques (such as lean, Six Sigma, lean, ERP, Service Oriented Architecture SOA, and other enablers [23]) and embraces parts of management, like Change management, Information Technology (IT) management, Project management and deals with suppliers, customers, employees, *etc....* [22]. BPM is a comprehensive approach to process improvement and it is focused on integrating, aligning, managing and measuring all the business processes of an organization. It is an inclusive approach of other approaches, it includes the application of all the other available methods (Process Excellence, Performance Improvement, Six Sigma, Business Process Reengineering, Lean, Business Process Engineering, Customer Expectation Management) [24]. BPM is the synthesis and extension of several technologies and techniques into a unified whole, a new foundation upon which to build sustainable competitive advantage [25]. Therefore our proposal lies in an integrated solution that embraces all the earlier proposals that allow Mass Customisation, while provides continuous improvement, customer satisfaction and constructive sustainability.

The rest of the paper is structured in the following way: In Section 2 a review and analysis about the state of art of the research proposals and related works is done; in Section 3 the research methodology is explained;, in Section 4 the problems and weaknesses associated with the current process are described; in Section 5 partial solutionsare presented and the integrated management model proposed; in Section 6 an use case implementation and validation is defined,and finally in Section 7, the main contributions and main conclusions of the research and the way forward for future research are presented.

2. The State of the Art in Mass Customisation Applied to Construction Industry

Nowadays, there are several proposals that suggest the implementation of Mass Customisation inside the construction industry. One of the most cited authors in the field, Ph.D. Masa Noguchi [26], highlights the importance of Mass Customisation in the industrialized housing as a marketing strategy to solve customer dissatisfaction inside the Asian real estate market. Noguchi focuses on resolving the customer dissatisfaction through various customisation options. By offering customers the opportunity to express their specific needs with a support system for decision making in customizing homes. Based on online information technologies, Noguchi's approach provides the necessary support, with the available solutions and the estimated costs of the choices. This option is only available for prefabricated homes, so what about the traditional construction? In this way, the author [11] also introduces the implementation of Mass Customisation in traditional construction projects, but this time the newness consists in the application of a communication system that allows users to take part in the customisation of their homes.

In the same way, other authors [27] suggest that customers are involved in their housing customisation through a support system for decision-making according to their preferences using a hybrid approach, that combines case-based reasoning (CBR) and genetic algorithm (GA). Through this system the possibility of customer dissatisfaction with the final product is reduced, solving the problems caused by inexperience for decision making cost-quality of the chosen option, or by discrepancies between expectations and final outcome of the house.

Also other researches [28] improve the customer participation by encouraging the exchange of information through an interactive models for customers that include BIM (Building Information Modelling) technology. With this approach they incentivize customer satisfaction while gaining the benefits of using BIM (Building Information Modelling) technology, as getting a better integration of data, fewer mistakes and inconsistencies in the projects. In addition to this, through the BIM system can be extracted 2D drawings and 3D models from any point of view, reducing time, work and mistakes that come from project modifications.

Although all these solutions improve customer satisfaction, giving them the opportunity to participate in the design process by customizing the house according to one's preferences, they provide a partial solution because customers are only involved in the design phase but they are not in the construction phase. So in none of them is proposed an integrated solution to the problems of efficiency in traditional construction.

Related to this another paper [29] presents an improvement, through the implementation of a production model that includes three sets, a design system, a construction system and a computer system, based on the codification of prefabricated building systems. The integration is achieved through a tool to visualize solutions and the automatic generation of the required information in the production phase. They present an integral system based on technology that speeds up the construction process while providing customer satisfaction, without increased cost. Nevertheless they offer a partial solution because it is only oriented to a prefabrication business model and actually not in traditional construction. In addition, it is only available for a modular design. Also the building system outlined in the paper follows a procedure not a customer oriented process. It is not process oriented and there is a delinking between process and technology.

As well, some authors [30] propose the use of technology to overcome the disadvantages arising from the Mass Customisation, such us increased manufacturing cost, fewer on-time deliveries, low supplier delivery performance, and increased order response time, increased material costs and reduction in product quality. To reduce these shortcomings and the bad influence on the construction costs they use information technology (IT) to give speed to the process, accessibility and exchange of information. By the use of a system called FIS—Finishing Information System, they improve the productivity in housing construction, minimizing the additional manpower due to the use of Mass Customisation, improving the efficiency of communication between providers and consumers, even if the process participants are geographically separated.

Although the proposal demonstrates the adequacy of Information Technology (IT), it is thus important to note that it is not focused on process management or the continuous improvement processes, neither are they measured. In addition to this there aren't defined actors and the functionalities are not established.

Despite all the literature reviewed for Mass Customisation applied to construction, the mostly analysed solutions focus their interest on the design phase. Thus, they do not do in the site execution phase. Also hardly any of them are centred on a technological solution. And the ones that take advantage of ICT's do not offer an integral management system to solve the problems that impede to achieve Mass Customisation in the field, such as the lack of business strategies that allow better efficiency in operations and people involved in them using technology holistically.

3. Research Methodology

Based on the analysis performed in the previous sections, it has been identified the need to propose a comprehensive model for the execution of the works oriented towards mass customisation. It allows achieving a strategic position in the market and satisfying customer needs by adding value to the product. At the same time, the necessary requirements to achieve this model were identified, obtained from observation, experience and analogy to other areas where similar problems have occurred. The synergy between the different management strategies, that are identified, together with ICTs will be the key factor to achieve these requirements.

To carry out the proposal it has been followed a research methodology based on business process management [31–33]. Process management is a strategy for structuring a complex process into a sequence of tasks understood as actions that transform inputs into some other output elements. This transformation must be aligned with previously defined goals, which are considered as strategic, to meet some needs or gaps identified in the environment. In this way, the defined process will achieve the object model of this work in a systematic way, selecting the most appropriate techniques and tools to meet the objectives and thus solving the problems identified. In the proposed methodology, each of the tasks identified represents a stage of the investigation, and these will have associated one or more scientific methods as described below.

The proposed research methodology is based on Eriksson-Penker formal notation [34]. It is an extension of UML for the representation of business processes that is characterized for being very intuitive and easily understood by all the stakeholders involved in the process. This is a notation that facilitates the understanding of the problem and its subsequent analysis to identify a solution according to the objectives.

Specifically, from the starting hypothesis Figure 1 shows the main process carried out in the research, where the input element (<<*input*>>) represents current models of mass construction. This <<*input*>> must be transformed through this design process into a building model oriented towards mass customization (<<*output*>>), meeting the requirements identified above and which now represent the strategic objectives that will guide the research process (<<goals>>).

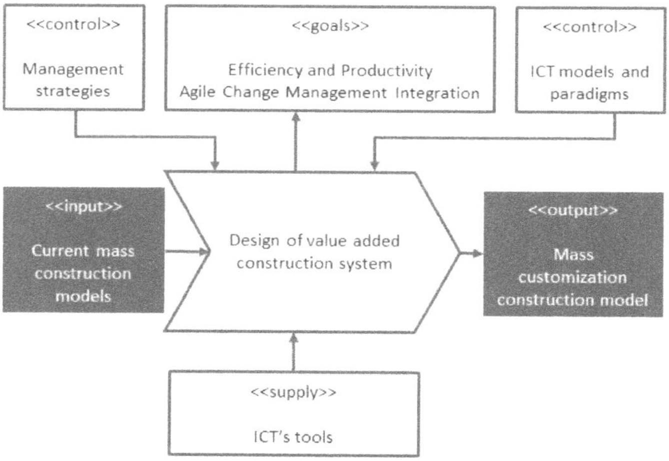

Figure 1. Modelling of value added design construction system process through Eriksson-Penker notation.

To perform the transformation to achieve the objectives established, must be identified the controllers (<<Control>>) and facilitators (<<supply>>) needed to guide this transformation. Particularly, these elements represent strategies, paradigms, techniques and technologies that will be integrated into our proposal. To facilitate its identification from a better comprehension of the problem, processes can be divided into sub processes or tasks. In our case a first division has been required as illustrated in Figure 2.

The resulting sub processes are, in first place, the sub process "identification of the construction site problems", focused on locate those general problems of mass construction current models that impede the achievement of the goals of the main process. In second place the sub process "partial solutions analysis", centered on the identification and analysis of solutions to the problems identified in the previous sub process. And finally, the sub process "integration of solutions" focused on obtaining an integral solution to the problem.

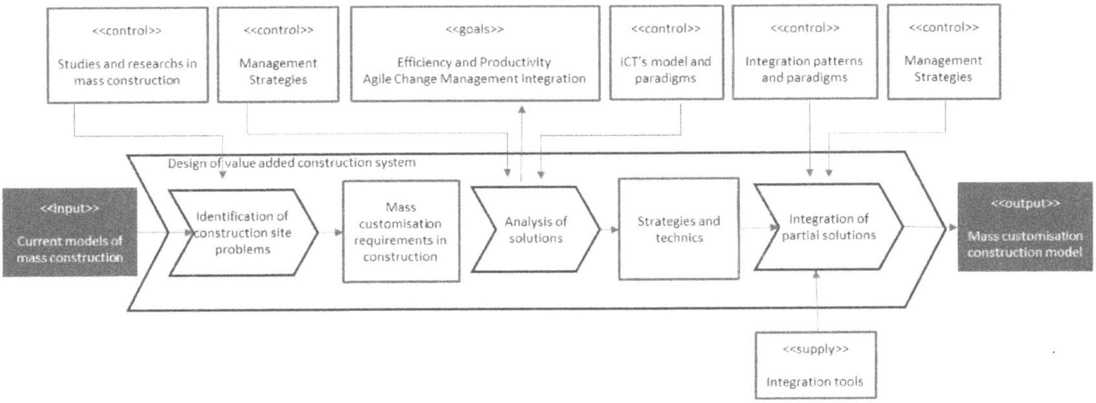

Figure 2. Modelling of value added design construction system sub processes through Eriksson-Penker notation.

In the following sections and using this methodology, will be developed each mentioned sub process, justifying the achievement of the general and particular goals and describing the actions performed to obtain the proposed model.

4. The Analysis of the Construction Site Problems

As we have specified before, one of the main barriers for adopting Mass Customisation as a competitive strategy in the sector is the increase in uncertainty, complexity, cost and time. These inconveniences are further compounded by the already low standards of efficiency and productivity that the sector is dragging in recent years. Compared with the traditional manufacturing industry, the construction sector has a 30% of less productivity levels [35] even though it sets up as a key sector within the framework of the Spanish economy. It reached at its peak the 11% of Spanish Gross Domestic Product GDP [36].

4.1. Identifying the Causes of Such Problems

These shortcomings in productivity and efficiency levels produce a significant increase in construction costs, due mainly to some causes that have their origin in the business organization such us obsolete management models, lack of business integration, improvisation in strategic decisions or search for short-term benefit. Some of them are a result of the real estate boom experienced by companies during the housing bubble [35]. Therefore, to be successful, the construction companies need to supplement their current short term approaches with more long term strategies [37,38] and move from traditional models management to advanced models management, based on the inter-and intra-enterprise integration [39] that respond to the new situation of the sector. Through better business integration and organizational structured models it is possible to take full advantage of functional synergies and reduce costs. With a systematic organization structure that takes into account the capabilities of stakeholders and the environment, avoiding making decisions by relying exclusively on intuition or experience.

Besides the business causes that promote low levels of productivity, mostly of the problems come from people who participate in the process. The construction works are characterized by the concurrence of variety of firms and participants. These problems appear mainly in two kind of agents: Contractors-subcontractors and workmanship [40]. On the one hand, they talk about the necessity of eliminating the adversity and competition atmosphere, that it is usually generated because of the war costs generated between subcontractors and their low involvement in the project. On the other hand, they point the low level of specialization and experience of the sector's manpower. Also they stipulate that the successful key is to become the adversity tendency between different parties, in adaptations and collaborations for mutual success, and to implement security and training oriented programs for all the workers. Related to the lack of formation also it can be included as human causes the unstructured traditional works prone to improvisation, without a systematized procedure [41]. Besides to this, they state the human causes that promote low productivity as lack of involvement, poor worker performance, low motivation, absenteeism, turnover and changes in the work teams, insufficient space to do their jobs efficiently, and the presence of obstacles in workspaces. There are also quite frequent incidents during the course of the construction works, such as project changes during execution, design errors and re-works made by the repetition of activities badly performed [42].

As all they have suggested, the human causes have a huge impact in productivity. Furthermore this situation worsens because of the heavy dependence of the building sector on human resources rather than adopting advanced or automated technology [43].

Although technology is one of the greatest allies of the construction companies to improve the competitiveness of their business, the specific software for business management does not have much presence in the sector. Indeed, the large majority of the construction micro-enterprises (97.9%) do not have ERP software tools [44]. Furthermore, the ones that use them show up a technological waste because the ERP package can hardly meet the needs of the construction industry[45] because they come from another sector. Due to its manufacturing origin, none of the construction management modules of commercial ERP packages provided by software suppliers have been found suitable for construction firms [46].

Then also, the main technological trend in the sector is the use of management software focused only on specific tasks such as project planning and monitoring, cost control, risk management, scheduling, *etc.* [47]. Also the whole process is divided into many isolated stages such as design, tender, construction, and maintenance, so the software applications used in each stage has led to the creation of "islands of automation", *i.e.*, incompatible hardware and software, because of the lack of attention given to the integration of these applications [48].

This fact favours the tightness of information, impeding information being transferred from one stage to another [49]. Besides this and related to the incompatibility problem mentioned earlier, there is a technological gap because of the lack of a standardised platform for information exchange that can facilitate the flow of information between incompatible hardware and software [50]. This technological gap promotes the existence of errors due to the inadequate, incomplete and outdated information that usually lead to delays and extra costs during the execution of a construction project [51]. About these extra time and cost of operations, in [52] is emphasised the current lack of discipline

in planning and execution tasks, due to the flow variation that increases operating cycle times. And in this sense, it makes worse the coordination of interdependent disciplines or crafts with the consequent unneeded inventory or lack of materials that produce incompletions tasks, and again, errors requiring work to be redone and re-handly. As well there are other factors that also cause missed deadlines and increased cost because of the uncertainty in the building construction, and the shortage of tools to address this uncertainty and these changes in a flexible manner. Change is inherent in construction works [53] due to some peculiarities such as complexity and uncertainty [54] resulting from one-of-a-kind nature of projects, the production site, and the temporary multi-organization [55].

4.2. The Classification of the Causes

Following the Leavitt [56] model classification the main factors involved in an organization and their interaction are distinguished. After that, it has been made a classification of the causes in order to help to identify the solutions. Then, all the previous causes that promote uncertainty, complexity, and deviations in cost and time, have been encompassed in four mainly groups of factors: Causes that have its origin in the structure or business organization, causes that have its origin in people who participates in the hole construction process, causes that have its origin in the applied technology and causes that have its origin in the tasks of the process.

Furthermore these causes are aggravated on account of the lack of integration and interaction, hampering the improvement of competitiveness in construction. This interdependence is shown in some researches [51,57] where is stablished the four interacting components that make up an organization (task, structure, people and technology) and how the change in one them inevitably affect the other three. This synergy implies that major improvements are made by house building companies by focusing on all the interrelated actions areas.

Therefore to reach the paradigm of Mass Customisation in building is necessary a comprehensive model that solves the requirements defined in Figure 3. To solve these causes is need competitive essential actions such as a change of mind, a shift toward process orientation and improved communications [51]. By encouraging better communication intra (inside organization level) and inter (customers-subcontractors and suppliers) companies with cohesive relationships based on trust, mutual commitment, understanding of each other's individual expectations and an open book culture (open exchange of information), organizations remove problems about information and the bottleneck that can lead to delays and extra costs. Also some key drivers of change to solve problems of cost and time are: Committed leadership, a focus on the customer, integrated processes and teams, a quality driven agenda and commitment to people. This last one also includes decent site conditions, care for the health and safety of the work force, including commitment to training for all participants in the process, involving everyone in sustained improvement and learning [58].

In this regard, some of the current new paradigms or technological solutions to achieve construction efficiency are only sub optimizing individual factors instead of optimizing the entire system [59]. Then, the expected benefits are not achieved because there is not a strategic vision in the organization or technology do not form an integral part of the organization business strategy [60]. Then, the appropriate way of implementing changes and innovation is enhancing the overall organizational performance [61], emphasizing a holistic model that includes innovation in business strategy, organization of work, technology, and people. So to take full advantage of functional synergies, changes cannot be done in isolation, it is necessary to develop a holistic system where the factors, structure, tasks, people and technology, are linked (Figure 4).

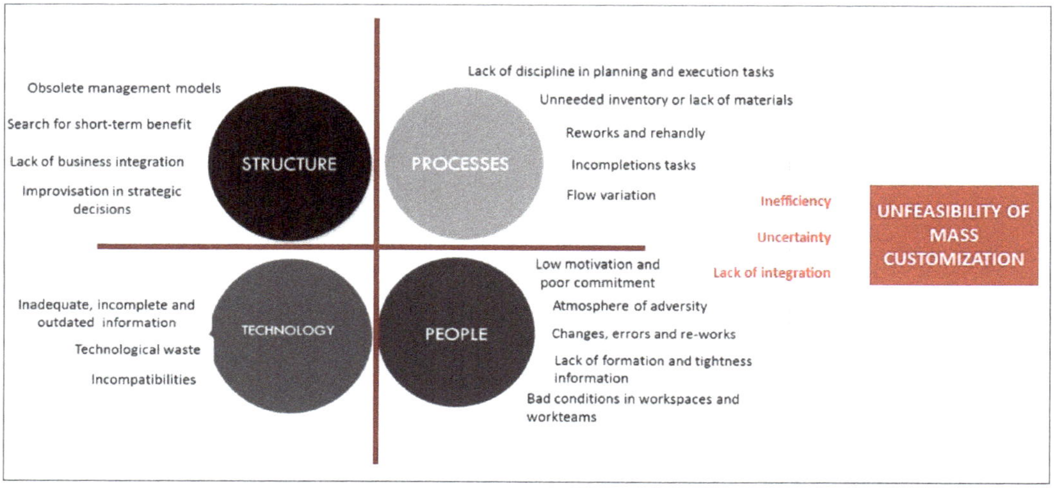

Figure 3. The Classification of Causes.

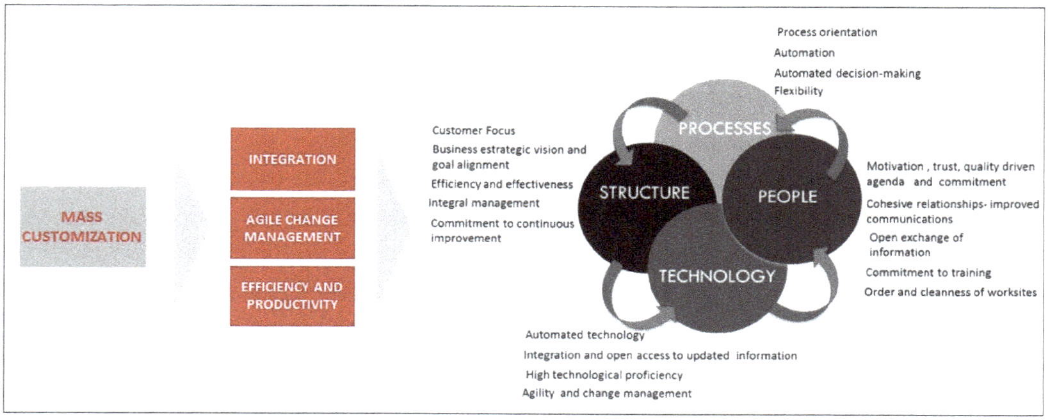

Figure 4. Requirements for Mass Customisation Feasibility.

5. An Integral Management Model for the Works Execution Oriented towards Mass Customisation

Within the construction field, there are different novel proposals that are focused on some of the objectives identified above. However, it comes to independent and partial solutions that do not offer an integral solution, which contains wholly the above objectives (Table 1). Therefore, a general and integral model it is proposed to achieve new paradigms, like Mass Customisation, in construction. The proposal is based on the integration of the above solutions, which offer partial solutions to the considered problem, together with other proposals that have proven successful in other sectors to solve the rest of problems that impede achieving Mass Customisation.

Table 1. Objectives Achieved with the Existing Techniques.

Existing techniques	Objectives Achieved with the Existing Techniques														
	Maximize customer value	**Standardization. Reduce variability**	**Minimize waste**	**Continuous improvement**	**Trust and commitment**	**Improved communications**	**Open exchange of information**	**Order and cleanness of worksites**	**Automated technology**	**Open access to updated information**	**Automated decision-making**	**Integration (not just technology)**	**Compatibility and high technological proficiency**	**Agility and change management**	**Flexibility**
Lean Construction	■	■	■	■	■	■	■	■							
Six Sigma	■	■		■	■		■			■					
ICT's									■	■	■				
BPM	■	■	■	■		■	■		■	■	■	■	■	■	■

One of the emerging mainstream approaches in the construction industry is the application of the Lean Construction philosophy, the new form of production management that comes from the manufacturing industry [14].

The aim of Lean Construction is to reconceptualise construction as flows to maximize customer value while minimizing waste, understanding waste [62] as any activity which absorbs resources but creates no value, such as mistakes which require rectification, processing steps which are not actually needed, movements of employees and transport of goods from one place to another without any purpose, groups of people in a downstream activity standing around waiting because an upstream activity has not delivered on time. Thus, Lean helps the firm to define customer value focusing on activities that add value, through standardizing the types of waste and offering best practices to remove waste. And therefore achieving, among other things, to reduce inventory and waiting times [63]. In addition, as it is established by Professor L. Koskela [54,55] other main focus of Lean

Construction is to increase output value through systematic consideration of customer requirements, to reduce the cycle time, to simplify by minimizing the number of steps, parts and linkages, to increase output flexibility, to increase process transparency, to focus control on the complete process, to build continuous improvement into the process, to balance flow improvement with conversion improvement and benchmark, and to reduce variability. Understanding as variability the construction uncertainties that produce late delivery of material and equipment, design errors, change orders, equipment breakdowns, tool malfunctions, improper crew utilization, labour strikes, environmental effects, accidents, and physical demands of work [64], to eliminate these variations and create workflow in a process is widely recognized the use of the LSS method [65]. Six Sigma is a statistical-based methodology that provides a structured framework to organize and implement strategic process improvement initiatives to attain reductions in process variability [66]. Within the construction industry, Six Sigma has been applied independently to improve the overall performance by reducing process variability in current construction operations [67]. Moreover it has been implemented as a quality initiative in traditional construction [68], increasing customer satisfaction and profitability by improving the quality of products [69]. Also it has been used for sustainability and quality improvement in prefabricated composite structures, applying Six Sigma to increase quality management while reducing the consumption of energy during construction, pollution, noise pollutions and waste [70]. Similarly, to reduce and eliminate waste in prefabricated residential construction, Lean principles and Six Sigma has been used by working in unison as an integrated model [71]. This combination of Lean tools and Six Sigma has been also applied in traditional construction, where each methodology complements the other, to eliminate waste and variations and create workflow [72,73]. Lean does not possess the tools to reduce variation and bring a process under statistical control, and in the same way, despite the fact Six Sigma offers a lot of gains alone the process would be slow and its costs will be too high. More specifically, lean benefit Six Sigma in the identification of waste providing the powerful value stream map tool that highlight waste and delays, while improve process speed or cycle time [74].

Despite the common factors and the synergy benefit, Lean mainly revolves about the relation between people and process but it does not on technology [75], in a similar way the use of IT in Six Sigma projects is often carry out with specific statistical software [76]. Therefore, despite the benefits obtained in the use of LSS, it is not resolved the gap in terms of integration with the other factors identified above technology, structure, people and structure.

Furthermore, the introduction of these paradigms requires a great effort of change in thought and culture, in most cases made up for a change agent to facilitate its immersion[30]. There are other issues that do not cover the integration and agility to manage change. ICT misguided (BPR) can be counterproductive for achieving integration and agility (bottom-up approach).

Moreover, ICT's are presented as an indispensable facilitators that can help to achieve mass customisation in construction, by addressing some of the goals that have not been resolved by earlier paradigms such as the process automation, the technology integration, the support in decision making, *etc.* [77].

In other traditional sectors, there are now some proposals to implement strategies such as mass customisation [31], by using a set of paradigms and technologies oriented to solve some of the main obstacles that hamper the application in the construction industry.

These paradigms and technologies are focused on getting flexible systems that allow adapting quickly the technology to the business processes of an organization, by proposing fully interoperable open systems (such as Service Oriented Architecture SOA, Web Services and Restful or Enterprise Service Bus) [78–80].

In addition, they consider the reuse of legacy systems like the ones used today in construction, the implementation of which has involved high cost and should not be discarded if not transformed into elements with the necessary functionality to support the processes of the organization.

Also the use of intelligent techniques based on semantic reasoning have brought great benefits to achieve mass customisation, enabling the automation of certain management steps like modelling (such as ontologies and associated techniques) [33].

BPM as the Backbone

To meet the stated goals, it is necessary to join these views or partial solutions because when the integration is done properly, the benefits provided individually will be increased [71,72,81]. For this it is necessary to achieve the integration of all the factors and resources involved in the execution of the work. And also to be able to reach an agile change management, allowing the immediate alignment between the strategic objectives, the defined business processes to achieve those objectives and the technologies that support the execution of the processes.

In this sense as the main novelty to cover these gaps, it is proposed the strategy Business Process Management (BPM) [82] to support the model which not only allows achieving those goals, but also enhancing the benefits obtained by the other proposals such as efficiency, productivity and agility (Figure 5).

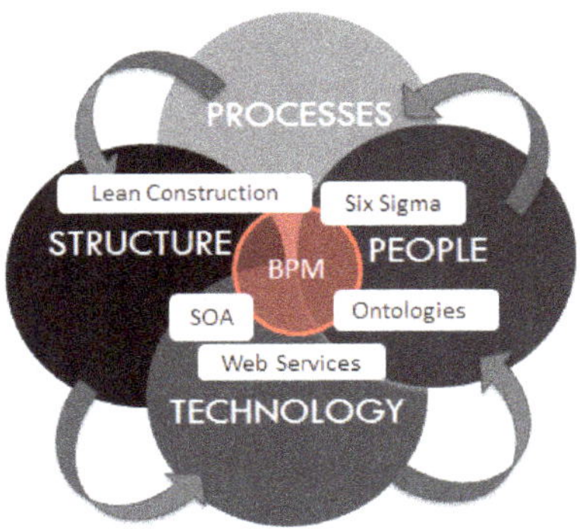

Figure 5. BPM as the Backbone of the Four Interacting Factors in Organizations.

BPM is a process-oriented business strategy focused on continuous improvement which includes IT as a fundamental element, unlike other strategies, filling the gap between business needs and IT capabilities [83], managing the complexity of the diverse inherited technology portfolio, creating transparency in the business environment and creating the agile link between business strategy and its execution [84]. All this is achieved in form of a continuous process management lifecycle [85], consisting of discovery, design, implementation, deployment, execution, interaction, control monitoring, analysis and optimization phases.

Associated to the BPM movement has emerged a software solution focused on process management that supports the principles and the BPM lifecycle model, allowing the integration of people, systems and suppliers, the automation and the processes collaboration. They are called the BPMS (Business Process Management System or Suite) [86]. The technological components linked to this system are shown in Figure 6 [25,82] and they are what give its key features and benefits: Monitoring and control the information of the key performance indicators process, to reduce costs by automating and improving processes in real time, to adapt processes to changes in an agile way, immediate deployment of process-centric applications, to carry out solutions that face up to the needs resulting from continuous change, and use the investments in existing IT.

Taking as reference the BPM paradigm, whose life cycle will be supported by a BPMS system, the benefits of the solutions presented individually will increase [66,71,81]. The presented model is structured according to the top-down strategy by providing two levels, the level of management and the level of resources respectively, which will be connected through the BPM system (Figure 4). This approach is important because to achieve the proposed objectives it is necessary a complete understanding of the processes before being automated.

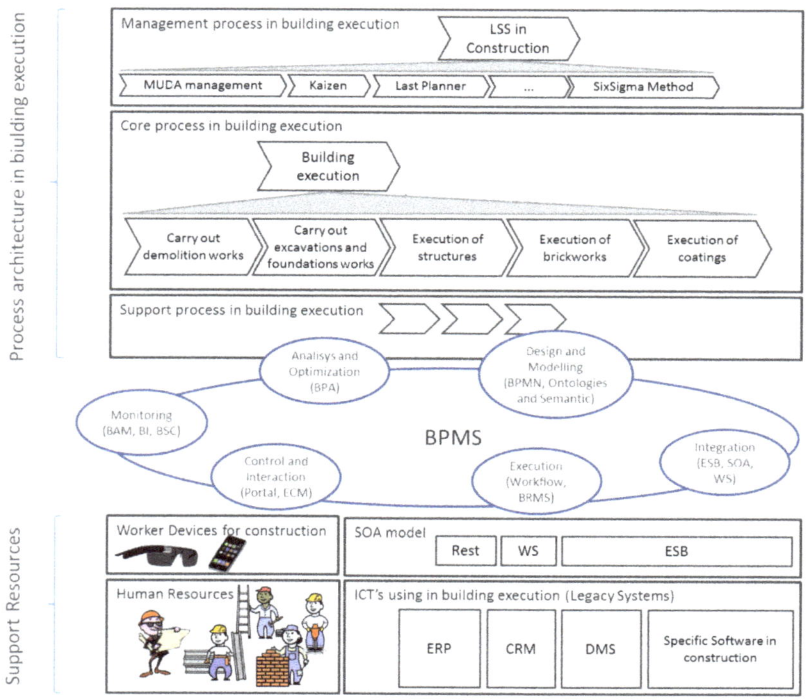

Figure 6. A Construction Management Framework for Mass Customisation in Traditional Building.

At the management level it is placed the work execution process architecture including process flow, the objectives to be achieved, the key performance indicators (KPIs) associated with the objectives of achievement measures, and the different stakeholders.

The BPMS includes tools for modelling and to include the elements defined in the architecture. The processes are modelled using BPMN standard notation, specifying in each process or task how to calculate KPIs and where to get the data, and associating the actors that should or may perform the different tasks to the BPMN model. BPMN is the standard notation for process modelling and that at any BPMS has a direct translation to the format that supports the processes execution, workflows owners or based on standards such as BPEL, BPEL4People and WS-HumanTask. Also BPMN is an understandable and intuitive notation based on flow charts, with a high degree of expressiveness that allows the process participants to understand how they should perform the tasks. The BPMS will support the processes included in this architecture that have been structured into three types, following any of the recommendations from some of the main authors [23,25,87] processes management, core and support processes.

A key aspect of the proposed model is that it has been included within the layer of management processes the ones related to Lean Construction philosophies, Six Sigma in construction or LSS in construction that are focused on the processes efficiency of the organization. Specifically, are implemented as processes and procedures the different techniques and principles associated with these disciplines, such as Kaizen, Last planner system, statistical techniques or any other analysis technique for continuous improvement.

One of the problems identified in the application of Lean and Six Sigma in the construction environment is the difficulty to assimilate the implementation of these philosophies, due to their high cultural dependence [30]. The automation of these principles and techniques with tools related to BPM will help to facilitate the inclusion. BPMS includes process execution engines and incorporates task managers that show the work to be done by each actor according to the running of the process flow, storing the relevant information and disposing of it in real time anytime and anywhere. Not only will this help to accelerate the process but also to know what to do, how, when, and how long (standardization of Lean) to each participant, both in terms of production processes and to apply continuous improvement techniques for the government of such processes.

For example, the site manager will be guided by the Last Planner process to manage daily in an agile manner the project execution, assigning tasks every day and even changing that assignment, appearing instantly to every member of the affected team the process to perform, how and when. If a construction's worker has allocated the paving execution of the houses on the first floor and each customer have chosen a different kind of pavement, in each case, it would appear the correct custom process that the worker should follow for put it in the more efficient manner, as well as the estimated time (tacktime in Lean). Finally, when he would have finished the task, would conclude the work indicating the completion of the task, and then, the time spent on it and the materials used would be recorded. This could lead to a request for material purchase depending on stock and needs, or a process improvement analysis if the time spent were far above the tacktime. Even, any worker could throw a proposal for improvement initiating the Kaizen process to analyse the proposal and carry it out.

In the layer of core processes y support processes will be located the associated production processes of the value chain in the building execution and the processes that support the achievement of them, respectively. These processes will be designed according to the principles and techniques of Lean Six Sigma and BPM as it is shown in some studies [24,81] applied to the construction peculiarities. For example, standard processes based on the main guidelines and recommendations in the building execution field must be created to achieve greater efficiency.

However, in contrast to what happened with the BPR paradigm, it is not about getting fixed and inflexible processes but rather the proposed model will include the variability among its principles, allowing in an agile way to adapt the processes to the specific objectives of the organization and its continuous change through composition techniques included in BPM and related technologies like SOA and Web Services [78–80].

A further key and innovative aspect of the proposed model is the use of ontologies and semantic reasoners to streamline the generation of process modelling for the work execution. This issue directly affects the customisation efficiency, now that depending on customers choices, the appropriate processes to build the custom home will be generated connecting sub processes as if it were a puzzle. These processes will guide the work of other staff to avoid errors. These processes will be executed over the process engine or BPMS workflow, but to execute certain tasks or sub processes they must interact with external information systems that were presented as functional isolated islands within an organization, performing functions inside the process of the work execution. These systems are located in the resource layer of the model (Figure 6) and they will be exposed following the principles of SOA paradigm using Web Services technologies or ESB infrastructures.

In this way is removed the stiffness of these systems and their functionalities, providing aspects such as reusability, interoperability and alignment with business processes that increase the efficiency of the model. And thus, to take advantage of the existing resources which have been a high cost for the organization. The BMPs include an integration module that allows these technologies to interoperate perfectly. In fact, although BPM and SOA have emerged as two independent approaches, today we have tested the benefit of the joint use of both philosophies. Another essential element in the model is the staff, which is directly involved in the process of building execution.

BPMS provides several technologies that can be integrated into these processes. Firstly, technologies such as portals or ECM will allow different roles to perform several tasks associated with the process managing or human tasks that are a part of the process itself, as the buyer of the material. For it, it is proposed in the model the use of appropriate devices for the site features, where the use of devices such as mobile devices or google glass can be great allies.

Finally, the BPMS includes a model that offers different technologies allowing the process performance measuring in real-time, and the variability of the results through the KPI's defined. For the first case, this is where may be used Lean Construction techniques that would imply an improvement of the process. These indicators can be analysed manually by the person responsible, and this imply to throw one of the continuous improvement processes or may generate the execution of other processes to correct automatically possible deviations, as the application of material required. Another example would indicate when certain tasks are being diverted from planned tacktime.

In the second case, the indicators would be generated using statistical methods Six Sigma to locate deviations in the result of processes, such as e.g., the execution of the interior woodwork. If abnormalities were detected, process would be launched immediately to analyse the causes and propose possible solutions. This would allow us to reduce the work performed during the realization of the processes of finished works.

6. Use Case, Implementation and Validation

The theory model obtained previously presents an ambitious proposal in a framework that serves as a reference for the integral management for the execution of the works orientated to the achievement of mass customisation. The development of this model is to obtain a realistic proposal will be realized in different self-contained phases that by themselves contribute to achieving the efficient objectives and productivity and agile change management and integration. In concrete, in the present article it has been developed one use case of which the objective is the implementation of some representative techniques and principles defined in the model which permit reaching the strategic goals proposed. This relation is shown in Figure 7.

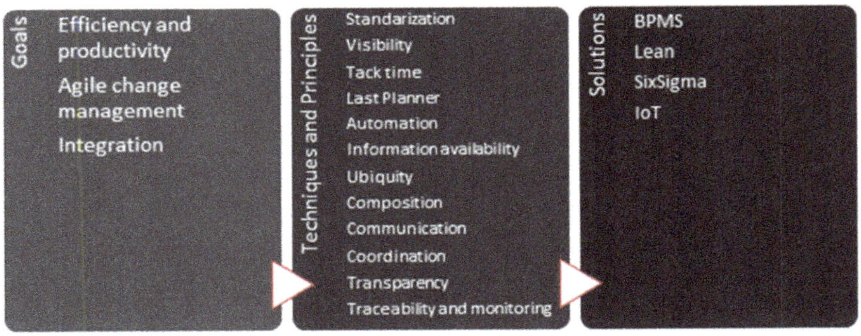

Figure 7. Relationship between objectives, principles and techniques and solutions.

In concrete, the use case includes two types of processes widely related, the process of weekly planning which implements the phase of the short-term planning of the Last Planner and which is located in the layer of management and the process of flooring, located in the layer core processes. Both processes have been chosen because of its repercussion in the mass customisation. The final customers could choose diverse types of floor and this includes that some customers could choose several types of floor for different rooms inside their houses. Also, this situation gets complicated and makes planning difficult because it requires a very high level of knowledge and control about the state of the execution of the works, concretely regarding the access of real time to this information because it exists too much variability and heterogeneity and human errors could increase considerably and therefore, increases the costs of execution and delays delivery deadlines laid out. On the other hand, it is possible that it is not known all of the proper procedures for each option of flooring or passing from one to another in a short period of time could derive in the inanition of some the steps.

Last Planner is a new technique of planning, used in Lean Construction that has broadly demonstrated its benefits in relation with efficiency and productivity. Nevertheless this is a complex process which implies a systematic methodology and a high level of coordination between the Construction Company and the different subcontractors. Also the information of planning carried out is realised in an extensive murals on the wall with the use of sticky's, which this information should be consulted in person. The election of the floor process has been because of different reasons. The flooring in a mass customisation model is a complex task due to the variability of the types of floor and therefore, to different procedures used for its execution.

The use of BPM as a management strategy to integrate other solutions has increased the benefits provided individually. In this sense, the use case was performed using the BPM open source Bonita BPM system (version 3.8). Firstly it has been done the modeling of the mentioned processes by using the standard modeling of business processes called BPMN (Figures 8 and 9).

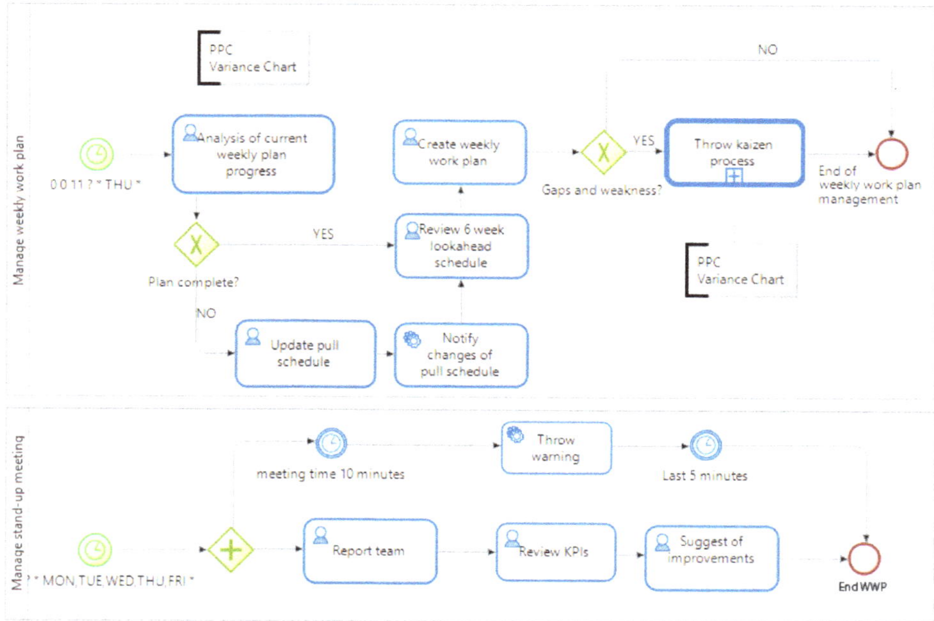

Figure 8. Weekly Work Plan BPMN model.

The process modeling has allowed us to achieve the following principles and their associated benefits:

- Standardization. It is a Lean principle that is increased with its realization by a standard modeling easily understandable by any role of the organization and based on a formal language that enables its automation. In both processes it will allow to stakeholders know what to do, how and by whom. Furthermore, in the case of the flooring, it has been done a redesign of the processes as a good practice based on the different techniques recommendations and the experience of several site managers.
- Another important aspect is the principle of processes composition in which BPMN is based. Modifications or link some processes with others is a very agile task that allows changing the behavior of a process in minutes. Any anomaly in the process could be rectified in a very short

space of time, reducing the number of errors by the same problem. Furthermore, each subcontractor could include its own modeled processes, facilitating the integration and incorporation of its system with the one of the builder. Another Lean principle that is facilitated thanks to the use of BPM.

- In this first phase of modeling, another important Lean principle that has allowed implement Bonita BPM, is the tack time. The tack time is the time required to perform a task depending on the time available for the entire process. To determine this time is essential to control more accurately the progress of the work execution. In fact, as we shall see, automatically we will get information in real time about what tasks are not being done on time, which can help the site manager to identify problems early.

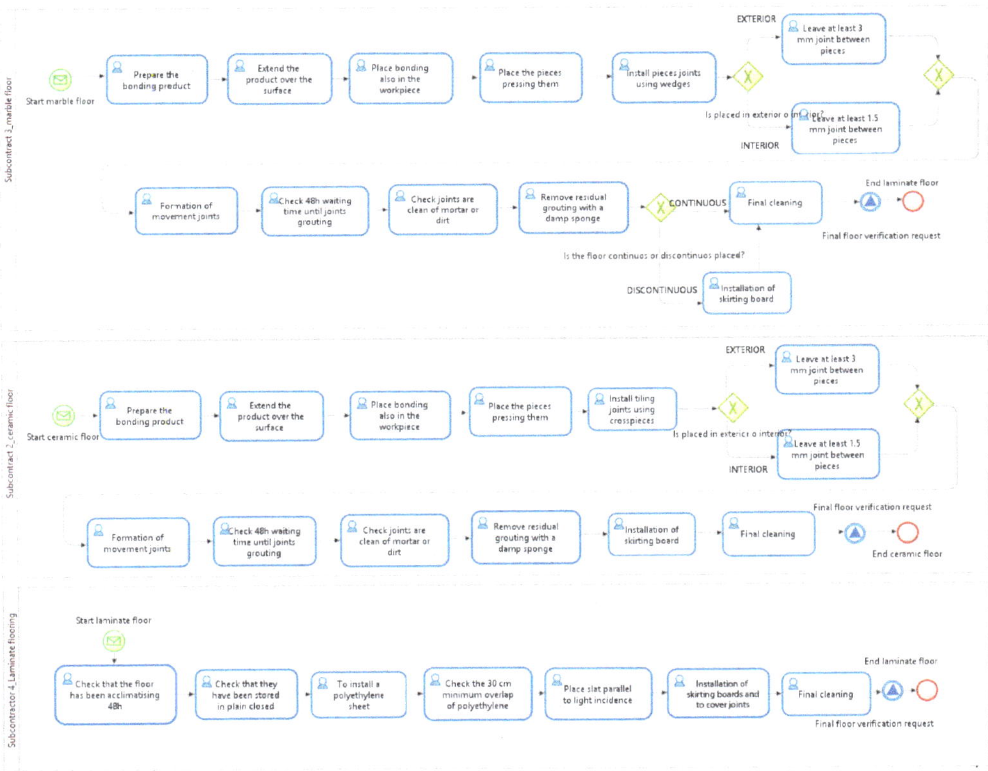

Figure 9. Flooring process BPMN model.

From the process modeling has been done the implementation of each of the tasks that compose the process, with particular emphasis on the needs of usability and accessibility in the built environment.

Firstly it has been developed a very flexible software tool that implements the Weekly Work Plan Process, till now performed manually. The entry point of this process is an event that is automatically launched every week, notifying everyone involved that must be met for the weekly schedule. This meeting could be virtual since they share a collaborative environment (Figure 10), involving considerable savings with respect to traditional methods. This planning is done starting from the tasks to be performed according to the 6 Weeks Look Ahead Schedule process. The environment implements a canvas and sticky's metaphor and includes a panel which allows using a simple drag and drop process to distribute tasks during the week, assign actors and specify the location where it must be carried out. At all times the user will have visual aids that reflect the state of the task.

Each task has an associated BPM process that establishes how to be executed and marks the execution time of the task from the tack time of each of the component tasks. Once planning is closed, the site manager can launch the process associated to each task, allowing the beginning of the execution of construction processes. In the tool these activities shall include a mark on the image that represents the task, enabling visualization at any time and from any location and device.

As shown in the Figure 10, for the same day of the week, Monday, it has been planned different types of floor to be executed. The execution of each process starts a series of events that would notify to the responsible of the tasks what to do together with the additional information for its management. These are human tasks, so to integrate the workers with the system devices have been used to facilitate interaction, offering a friendly, usable and simple interface. In this case we have done a prototype using platforms such us a smartwatch and a smartphone, as is shown in Figure 11. As the operator is carrying out the tasks, the processing sequence is advanced. The task management is very simple, is focused on reporting the task to be done, to execute the basic actions of it (start, stop, validation, *etc.*...).

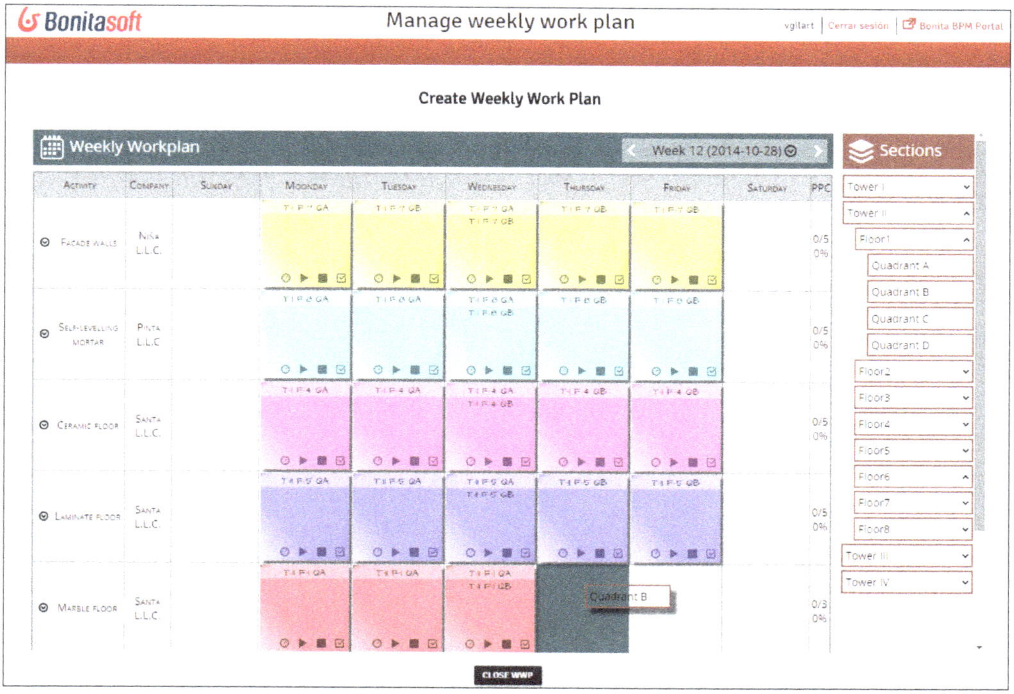

Figure 10. Screenshot of Weekly Work Plan application embedded in Bonita BPM Plannig phase.

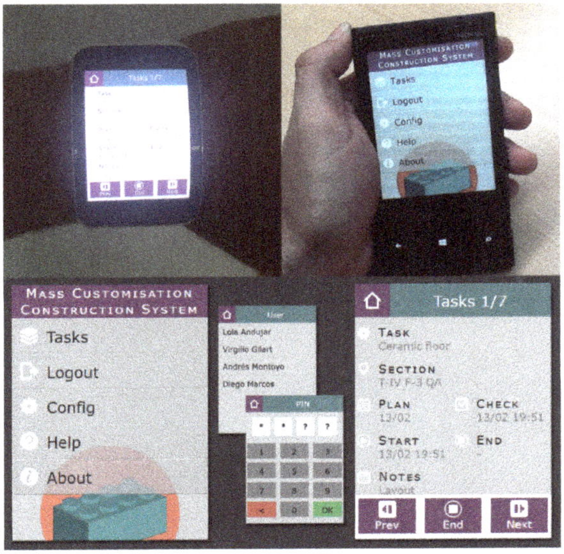

Figure 11. User agent for system interaction. Smartwatch and smartphone versions and interface screenshots.

Once a task is closed, the system notifies the next task to the appropriate worker. When the closure of a task involves the completion of a process, such action will be reflected in the application the Weekly Work Plan, showing the image of the task with a cross over it (Figure 12). Before starting a task, an operator can validate that the above task has been completely finished. This action involves, in the terminology of the weekly work plan, a double strikethrough on the visual representation of the task and updating the Percent Plan Completed (PPC) (Figure 12).

The implementation involves achieving the following principles, as is shown above in Figure 7, and its associated benefits to achieve the objectives of the proposal:

Automation. This is one of the BPM fundamentals, which allows the organization to automate the implementation, monitoring and control of its processes through the orchestration and integration of Processes, systems, people and information to create uniform, streamlined, and repeatable Business Processes. This automation is based on a process-centric approach where automation is performed on a redesign of the process based on best practices. Its automation implies an increase in performance in the process execution and therefore greater efficiency and productivity. Furthermore, the use of everyday devices based on IoT paradigm, in our case connected via Rest Services to the process facilitates the integration of human people in the same, allowing interaction and improving the efficiency and integration. Finally, it has been automated through events the notifications of tasks and actions to guide workers in the development of their activities. All this has led to improve the understanding and learning of the methodologies, techniques and philosophies implemented. Information available in real time and ubiquity. The proposal allows access to the state of the building process at any time, allowing each member involved to have complete information about it. Furthermore, such information can be consulted from any location and computer devices (pc's, laptops, tables, smartphones and wareables). This means saves on the time execution and therefore cost, also reductions of errors due to lack of information.

Figure 12. Screenshot of Weekly Work Plan application embedded in Bonita BPM. Traceability phase.

Communication, coordination and transparency. As was mentioned before, both the general framework of Bonita BPM and the applications developed on it, have been implemented in a collaborative work environment which together with the above principles of availability and ubiquity, enable synchronization and coordination of all stakeholders, which achieves the objectives of efficiency and productivity and integration. Traceability and monitoring. On one hand, the implementation of the Weekly Work Plan application allows for a visual feedback of the coarse-grained tasks. The automation of processes in Bonita BPM provides us full traceability and monitoring on the execution and the status of tasks, the time taken for completion, the person responsible who performed it and the task information that has been added. The level of detail provided is fine grained which will allow the site manager and work teams more precise control about each task. In this way it is possible to detect in real time the deviations of daily, and weekly planning and longer term planning the tack time defined (Figure 13). In fact, the system notifies the responsible any deviation on the tack time. All this information can be accessed from anywhere with different devices representing a complete knowledge about the execution of the work that influences in productivity, efficiency, and together with the composition, agile change management.

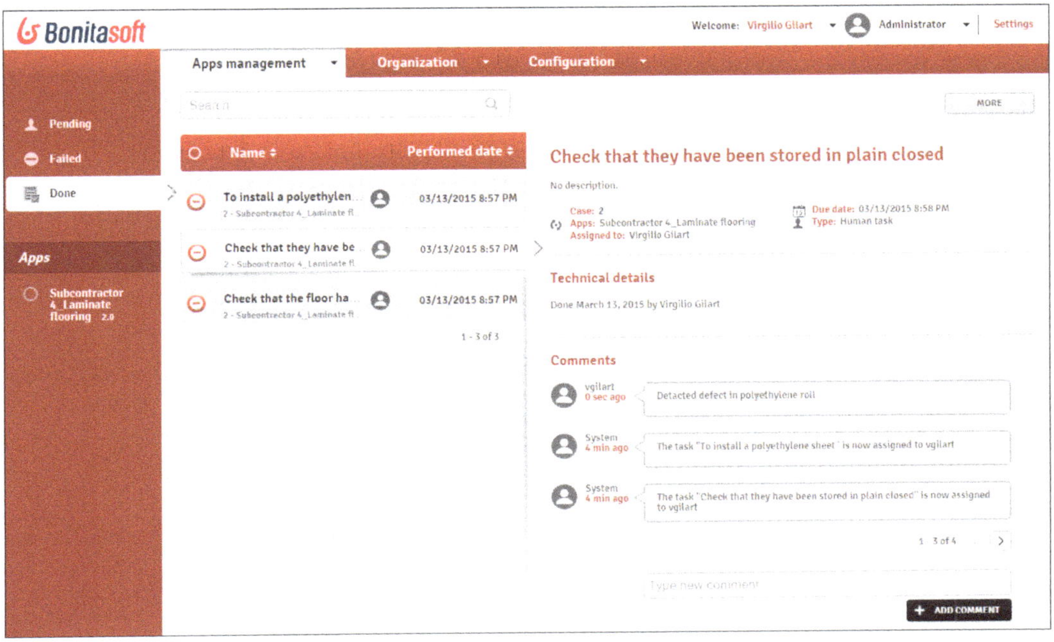

Figure 13. Bonita portal screenshot. Traceability and monitoring of flooring process.

7. Conclusions, Contributions and Future Work

In this paper we have presented a theoretical model for managing the execution of the work aimed at achieving carry out mass customization paradigm in traditional construction.

We have implemented a prototype model which has allowed showing its viability and has highlighted how the implementation of a representative set of techniques achieve the objectives of mass customization such as efficiency, productivity, agility and change management integration.

The main contributions of this work can be summarized in:

- It is the first time that BPM is applied to the field of construction in general, and the execution of the works, in particular. BPM allows achieving the principles of standardization, automation, agility, flexibility, integration. On the other hand, it is a generic framework that allows brings together and promotes the benefits of other strategies and to align management philosophies to align with information technology.
- It has been done an integral management model for mass customization in the execution of the work, particularly for traditional construction from the current management models.
- It has been carried out an integration of Lean and BPM paradigms; in particular, it has been automated using Bonita BPM a part of the Last Planner process (Weekly Work Plan). Until now there were some theoretical proposals offering the synergy benefits of these strategies, but none of them are focused on the specific area of construction.
- It has been incorporated everyday items based on the paradigm IoT to facilitate the integration of the construction actors in the process automation. Due to the characteristics of

the work environment, have been selected devices such as wareables with transparency features, usability, flexibility and accessibility.
- A redesign and standardization of the flooring process has been presented through the BPMN modelling language.

Currently, we are working on the implementation of new processes and modules of the proposed model, including Six Sigma strategy, in order to include them in the present approach.

Over this implementation, as future lines of work, we are going to develop an ontology of the building execution standard core processes, allowing the dynamic and automatic generation of the full process from customer's choices. In addition, this study is considered the first step to derive a global model that integrates the entire value chain for the construction, from the design stage to the delivery of the property.

Acknowledgments

This work was supported in part by the Vicerrectorado de Investigación, Desarrollo e Innovación, University of Alicante and by the Conselleria d'Educació, Cultura i Esport, Generalitat Valenciana.

Author Contributions

Virgilio Gilart-Iglesias and Andrés Montoyo are the Directors of the research in which all authors have contributed. María Dolores Andújar-Montoya has made the conceptualization and the problem definition, and together with Virgilio Gilart-Iglesias and Andrés Montoyo all they have developed the proposed model. Likewise, the designing of the system and its validation has been carried out by Virgilio Gilart-Iglesias, Diego Marcos-Jorquera and Maria Dolores Andújar-Montoya. The four authors wrote the paper, revised, read and approved the final manuscript.

Conflicts of Interest

The authors declare no conflict of interest.

References

1. Bernard, A.; Daaboul, J.; Laroche, F.; da Cunha, C. Mass customisation as a competitive factor for sustainability. In *Enabling Manufacturing Competitiveness and Economic Sustainability,* Proceedings of the 4th International Conference on Changeable, Agile, Reconfigurable and Virtual Production, Montreal, QC, Canada, 2–5 October, 2011; Springer: Montreal, QC, Canada, 2012; pp. 18–25.
2. Pine, B.J. *Mass Customization: The New Frontier in Business Competition*; Harvard Business School Press: Boston, MA, USA, 1999.
3. Tseng, M.M.; Hu, S.J. Mass customization. In *Cirp Encyclopedia of Production Engineering*; Laperrière, L., Reinhart, G., Eds.; Springer: Berlin, Germany, 2014; pp. 836–843.
4. Tseng, M.M.; Jiao, J.; Merchant, M.E. Design for mass customization. *CIRP Ann.-Manuf. Technol.* **1996**, *45*, 153–156.

5. Confederación Nacional de la Construcción (CNC). *Informe de Evolución del Sector de la Construcción*; National Confederation of Construction: Madrid, Spain, 2011; pp. 17–46.
6. Gilmore, J.H.; Pine, B.J. *Authenticity: What Consumers Really Want*; Harvard Business School Press: Boston, MA, USA, 2007; Volume 1.
7. Barlow, J. From craft production to mass customisation? Customer focused approaches to house building. In *From Craft Production to Mass Customisation? Customer Focused Approaches to House Building*, Proceedings of IGLC 6th Annual Conference, Sao Paulo, Brazil, 13–15 August 1998; Citeseer: Sao Paulo, Brazil, 1998.
8. Nambiar, A.N. Mass customization: Where do we go from here? In *Mass Customization: Where do We Go from Here?* Proceedings of the World Congress on Engineering WCE, London, UK, 1–3 July 2009; pp. 1–3.
9. Piller, F.T.; Moeslein, K. From economies of scale towards economies of customer integration: Value creation in mass customization based electronic commerce. In Proceedings of the 15th Bled Electronic Commerce Conference e-Reality: Constructing the e-Economy, Bled, Slovenia, 17–19 June 2002.
10. Radder, L.; Louw, L. Mass customization and mass production. *The TQM Mag.* **1999**, *11*, 35–40.
11. Noguchi, M.; Hernàndez-Velasco, C.R. A 'mass custom design' approach to upgrading conventional housing development in Mexico. *Habitat Int.* **2005**, *29*, 325–336.
12. Noguchi, M. The "mass Custom Design" Approach to the Delivery of Quality Affordable Homes. Available online: http://www.masscustomhome.com/ (accessed on 8 November 2014).
13. Cuperus, Y. Mass customization in housing, an open building/lean construction study. In Proceedings of Dense Living Urban Structures International Conference on Open Building, Hong Kong, China, 23–26 October 2003; pp. 1–13.
14. Aziz, R.F.; Hafez, S.M. Applying lean thinking in construction and performance improvement. *Alexandria Eng. J.* **2013**, *52*, 679–695.
15. Porter, M.E. *Competitive Advantage of Nations: Creating and Sustaining Superior Performance*; Simon and Schuster: New York, NY, USA, 2011.
16. Almeida, J.C.; Salazar, G.F. Strategic Issues in Lean Construction. In Proceedings of the 11th Annual Conference of the International Group for Lean Construction, Blacksburg, VA, USA, 22–24 July 2003; pp. 1–10.
17. Ahuja, V.; Yang, J.; Shankar, R. Strategic use of ict by construction organisations-requirements and implementation issues. *Bus. Process Manag. J.* **2006**, *15*, 968–989.
18. Rausch, P.; Stumpf, M. Linking the operational, tactical and strategic levels by means of cpm: An example in the construction industry. In *Business Intelligence and Performance Management*; Rausch, P., Sheta, A.F., Ayesh, A., Eds.; Springer: London, UK, 2013; pp. 27–42.

19. Cork, R. Is Construction Ready for Project Based Erp? Available online: http://download.ifsworld.com/home/if1/page_480/is_construction_ready_for_project_based_erp_yet.html (accessed on 24 April 2015).
20. Chung, B.; Skibniewski, M.J.; Kwak, Y.H. Developing erp systems success model for the construction industry. *J. Construct. Eng. Manag.* **2009**, *135*, 207–216.
21. Looy, A.V. *Business Process Maturity. A Comparative Study on A Sample of Business Process Maturity Models*; Springer International Publishing: Vaduz, Liechtenstein, 2014; Volume 1, p. 87.
22. Willaert, P.; van den Bergh, J.; Willems, J.; Deschoolmeester, D. The process-oriented organisation: A holistic view developing a framework for business process orientation maturity. In *Business Process Management*; Springer: Berlin, Germany, 2007; pp. 1–15.
23. Harmon, P. *Business Process Change: A Guide for Business Managers and Bpm and Six Sigma Professionals*; Elsevier/Morgan Kaufmann: Burlington, MA, USA, 2010.
24. Ohtonen, J. Combining all Bpm Methods–is that Possible? Available online: http://www.bpmleader.com/2012/07/02/combining-all-bpm-methods-%E2%80%93-is-that-possible/ (accessed on 2 October 2014).
25. Smith, H.; Fingar, P. *Business Process Management (BPM): The Third Wave*; Meghan-Kiffer Press: Tampa, FL, USA, 2003.
26. Noguchi, M. The effect of the quality-oriented production approach on the delivery of prefabricated homes in Japan. *J. Hous. Built Environ.* **2003**, *18*, 353–364.
27. Juan, Y.-K.; Shih, S.-G.; Perng, Y.-H. Decision support for housing customization: A hybrid approach using case-based reasoning and genetic algorithm. *Expert Syst. Appl.* **2006**, *31*, 83–93.
28. Lee, S.; Ha, M. Customer interactive building information modeling for apartment unit design. *Autom. Construct.* **2013**, *35*, 424–430.
29. Benros, D.; Duarte, J. An integrated system for providing mass customized housing. *Autom. Construct.* **2009**, *18*, 310–320.
30. Shin, Y.; An, S.-H.; Cho, H.-H.; Kim, G.-H.; Kang, K.-I. Application of information technology for mass customization in the housing construction industry in Korea. *Autom. Construct.* **2008**, *17*, 831–838.
31. Gilart-Iglesias, V. Metodología Para la Gestión Integral de los Procesos de Producción: Modelado de la Maquinaria Industrial Como un Sistema de Gestión de Procesos de Negocio. Ph.D. Thesis, Polytechnic University College, University of Alicante, Alicante, Spain, 2010.
32. Gilart-Iglesias, V.; Maciá-Pérez, F.; Marcos-Jorquera, D.; Mora-Gimeno, F.J. Industrial machines as a service: Modelling industrial machinery processes. In Proceedings of the 2007 5th IEEE International Conference on Industrial Informatics, Wien, Austria, 23–27 June 2007; pp. 737–742.
33. Ferrándiz Colmeiro, A.; Maciá Pérez, F.; Gilart Iglesias, V.; Berná Martínez, J.V.; Gea Martínez, J. Automatización del modelado y gestión de procesos de fabricación dirigida por semántica. In Proceedings of VI Jornadas para el Desarrollo de Grandes Aplicaciones de Red (JDARE'09), Alicante, Spain, 2009; pp. 181–198.

34. Penker, M.; Eriksson, H.-E. Business Modeling With UML: Business Patterns at Work. Available online: http://dsc.ufcg.edu.br/~sampaio/Livros/Wiley-Business-Modeling-with-UML-Business-Patterns-at-Work.pdf (accessed on 24 April 2015).
35. ISEA. La Innovación, una Prioridad Para el Sector de la Construcción. Available online: http://www.iseamcc.net/news/revista/revista-isea-03.pdf (accessed on 13 October 2014).
36. Cuadrado-Roura, J. *El sector de la Construcción en España: Análisis, Perspectivas y Propuestas*; Cuadernos del Colegio Libre de Eméritos: Madrid, Spain, 2010. Available online: http://www.colegiodeemeritos.es/docs/repositorio/es_ES/Cuadernos_del_Colegio/el_sector_construccion_en_espana_%28final%29.pdf (accessed on 13 October 2014).
37. Bakar, A.H.A.; Tufail, M.A.; Yusof, M.N.; Pinang, W.M.; Virgiyanti, W. Implementation of strategic management practices in the malaysian construction industry. *Pak. J. Commer. Soc. Sci.* **2011**, *5*, 140–154.
38. Price, A.; Newson, E. Strategic management: Consideration of paradoxes, processes, and associated concepts as applied to construction. *J. Manag. Eng.* **2003**, *19*, 183–192.
39. Vicedo, J.C.; Pérez, A.H.; Esteban, F.C.L. Análisis de adaptación al sector de la construcción de estructuras organizativas y de producción existentes en el sector del automóvil. In Proceedings of V Congreso de Ingeniería de Organización: Valladolid-Burgos, Spain, 4–5 September 2003.
40. Martín, A.I.F.; Frías, V.G.; Romero, B.P. La cadena de suministro en proyectos de construcción. In Proceedings of the II International Conference on Industrial Engineering and Industrial Management, Burgos, Spain, 3–5 September 2008; pp. 1715–1724.
41. Arcudia Abad, C.E.; Solís Carcaño, R.G.; Baeza Pereyra, J.R. Determinación de los factores que afectan la productividad de la mano de obra de la construcción. *Ingeniería* **2004**, *8*, 145–154.
42. Giménez Palavicini, Z.; Suárez Isea, C. Diagnóstico de la gestión de la construcción e implementación de la constructabilidad en empresas de obras civiles. *Rev. Ingenier. Constr.* **2008**, *23*, 4–17.
43. Chiang, Y.-H.; Tang, B.-S. “Submarines don’t leak, why do buildings?” Building quality, technological impediment and organization of the building industry in hong kong. *Habitat Int.* **2003**, *27*, 1–17.
44. Ontsi, F.Y.E. Análisis Sectorial de Implantación de las Tic en la Pyme Española. Available online: http://www.ipyme.org/Publicaciones/InformePyme2013.pdf (accessed on 17 October 2014).
45. Shi, J.J.; Halpin, D.W. Enterprise resource planning for construction business management. *J. Construct. Eng. Manag.* **2003**, *129*, 214–221.
46. Yang, J.-B.; Wu, C.-T.; Tsai, C.-H. Selection of an erp system for a construction firm in Taiwan: A case study. *Autom. Construct.* **2007**, *16*, 787–796.
47. Forcada Matheu, N. Life Cycle Document Management System for Construction. Ph.D. Thesis, Polytechnic University of Catalonia, University of Catalonia, Catalonia, Spain, 2005.
48. Cembellín, B.H. Herramientas informáticas para construir y fabricar. *Técnica Ind.* **2009**, *283*, 36–39.

49. Alshawi, M.; Faraj, I. Integrated construction environments: Technology and implementation. *Constr. Innov.* **2002**, *2*, 33–51.
50. Alshawi, M.; Ingirige, B. Web-enabled project management: An emerging paradigm in construction. *Autom. Construct.* **2003**, *12*, 349–364.
51. Hong-Minh, S.; Barker, R.; Naim, M. Identifying supply chain solutions in the uk house building sector. *Eur. J. Purch. Supply Manag.* **2001**, *7*, 49–59.
52. Ballard, G.; Howell, G. Implementing Lean Construction: Stabilizing Work Flow. Available online: http://www.leanconstruction.dk/media/18181/Implementing_Lean_Construction__Stabilizing_Work_Flow_.pdf (accessed on 24 April 2015).
53. Smith, N.J.; Merna, T.; Jobling, P. *Managing risk in Construction Projects*; John Wiley & Sons: Oxford, UK, 2013.
54. Koskela, L. *Application of the New Production Philosophy to Construction*; Technical Report No. 72; Stanford University: Stanford, CA, USA, 1992.
55. Koskela, L. Lean Production in Construction. Available online: http://www.iaarc.org/publications/fulltext/Lean_production_in_construction.PDF (accessed on 24 April 2015).
56. Leavitt, H.J. Applied organizational change in industry: Structural, technical and human approaches. In *New Perspective in Organization Research*; Cooper, W.W., Leavitt, H.J., Shelly, M.W., II, Eds.; Wiley: New York, NY, USA, 1964; Volume 55, p. 71.
57. Orozco, F.A.; Serpell, A.F.; Molenaar, K.R.; Forcael, E. Modeling competitiveness factors and indexes for construction companies: Findings of Chile. *J. Construct. Eng. Manag.* **2011**, *140,* 1–13.
58. Egan, S. J.; Williams, D. Rethinking Construction: The Report of the Construction Task Force. Available online: http://constructingexcellence.org.uk/wp-content/uploads/2014/10/rethinking_construction_report.pdf (accessed on 28 September 2014).
59. Newman, M.; Zhao, Y. The process of enterprise resource planning implementation and business process re-engineering: Tales from two chinese small and medium-sized enterprises. *Inform. Syst. J.* **2008**, *18*, 405–426.
60. Love, P.E.; Irani, Z. An exploratory study of information technology evaluation and benefits management practices of SMEs in the construction industry. *Inf. Manag.* **2004**, *42*, 227–242.
61. Sexton, M.; Barrett, P.; Miozzo, M.; Wharton, A.; Leho, E.; Hughes, W. Innovation in small construction firms: Is it just a frame of mind? In Proceedings of 17th Annual ARCOM Conference, Salford, UK, 5–7 September 2001.
62. Womack, J.P.; Jones, D.T. *Lean Thinking: Banish Waste and Create Wealth in Your Corporation*; Simon and Schuster: New York, NY, USA, 2010.
63. Howell, G.A. What is lean construction-1999. In Proceedings of Seventh Annual Conference of the International Group for Lean Construction, IGLC-7, Berkeley, CA, USA, 26–28 July 1999; pp. 1–10.
64. Abdelhamid, T.S.; Everett, J.G. Physical demands of construction work: A source of workflow unreliability. In Proceedings of the 10th Conference of International Group for Lean Construction IGLC-10, Gramado, Brazil, 6–8 August 2002.

65. Oguz, C.; Kim, Y.-W.; Hutchison, J.; Han, S. Implementing lean six sigma: A case study in concrete panel production. In Proceedings of the 20th Annual Conference of the International Group for Lean Construction IGLC-20, San Diego, CA, USA, 18–20 July 2012
66. Abdelhamid, T.S. Six Sigma in Lean Construction Systems: Opportunities and Challenges. In Proceedings of the Eleventh Annual Conference of the International Group for Lean Construction IGLC-11, Blacksburg, VA, USA, 22–24 July 2003.
67. Han, S.H.; Chae, M.J.; Im, K.S.; Ryu, H.D. Six sigma-based approach to improve performance in construction operations. *J. Manag. Eng.* **2008**, *24*, 21–31.
68. Pheng, L.; Hui, M. Implementing and applying six sigma in construction. *J. Construct. Eng. Manag.* **2004**, *130*, 482–489.
69. Tutesigensi, A.; Pleim, V. Title of Presentation. In *Why Small and Medium Construction Enterprises do not Employ Six Sigma*, Proceedings of 24th Annual ARCOM Conference, Cardiff, UK, 1–3 September 2008; pp. 267–276.
70. Tchidi, M.F.; He, Z.; Li, Y.B. Process and quality improvement using six sigma in construction industry. *J. Civ. Eng. Manag.* **2012**, *18*, 158–172.
71. Shan, Y.-H.; Li, Z.-F. Integration and Application of Lean Principles and Six Sigma in Residential Construction. In Proceedings of 2012 3rd International Asia Conference on Industrial Engineering and Management Innovation. Available online: http://link.springer.com/chapter/10.1007/978-3-642-33012-4_25 (accessed on 25 April 2015).
72. Banawi, A.; Bilec, M.M. A framework to improve construction processes: Integrating lean, green and six sigma. *Int. J. Constr. Manag.* **2014**, *14*, 45–55.
73. Sunil, V.D.; Sharad, V.D. Minimising waste in construction by using lean six sigma principles. *Int. IJCIET* **2013**, *4*, 1–8.
74. George, M.L.; George, M. *Lean Six Sigma for Service*; McGraw-Hill: New York, NY, USA, 2003.
75. Visser, B. Lean Principles in Case Management. Master's Thesis, Radboud Universiteit, Nijmegen, The Netherland, 2009.
76. Chiarini, A. Discussion and comparison about the common characteristics. In *From Total Quality Control to Lean Six Sigma*; Springer: Milan, Italy, 2012; pp. 47–51.
77. Chen, Q.; Reichard, G.; Beliveau, Y. Interface management—A facilitator of lean construction and agile project management. Proceedings of the 15th Annual Conference of the IGLC 15, 18–20 July 2007, East Lansing, MI, USA; pp. 57–66.
78. Erl, T. *Service-Oriented Architecture: Concepts, Technology, and Design*; Pearson Education Crwafordsville: Indiana, IN, USA, 2005.
79. Erl, T. *Soa Design Patterns*; Pearson Education: Boston, MA, USA, 2008.
80. Marks, E.A.; Bell, M. *Service Oriented Architecture (SOA): A Planning and Implementation Guide for Business and Technology*; John Wiley & Sons: Hoboken, NJ, USA, 2006.
81. Morris, D.; Field, G. BPM, Lean and Six Sigma Better Together the Whole is Greater than the Sum of the Parts. Available online: http://c.ymcdn.com/sites/www.abpmp.org/resource/resmgr/Docs/news_events_bpm_lean_6.pdf (accessed on 25 April 2015).

82. Jeston, J.; Nelis, J. *Business Process Management*; Elsevier Butterworth Heinemann: Burlington, MA, USA, 2014.
83. Woodley, T.; Gagnon, S. BPM and SOA: Synergies and challenges. In *Web Information Systems Engineering—Wise 2005. 6th International Conference on Web Information Systems Engineering*; Springer: New York, NY, USA, 2005; pp. 679–688.
84. Kirchmer, M. *High Performance through Process Excellence: High Performance through Process Excellence*; Springer-Verlag: Berlin, Germany, 2009.
85. Zur Muehlen, M.; Ho, D.T.-Y. Risk management in the bpm lifecycle. In *Business Process Management BPM 2005 Workshops*; Springer-Verlag: Berlin, Germany, 2006, pp. 454–466.
86. Chang, J.F. *Business Process Management Systems: Strategy and Implementation*; Auerbach Publications: Boca Raton, FL, USA, 2005.
87. Barros, O. Business processes architecture and design. *Bus. Process Trend* **2007**, *1*, 1–28.

Redesigning Terraced Social Housing in the UK for Flexibility Using Building Energy Simulation with Consideration of Passive Design

Hasim Altan, Nicola Gasperini, Sam Moshaver and Antonio Frattari

Abstract: A chosen case study house forms the basis of this paper, which is a pilot energy-efficient social housing project, completed by one of the largest housing developers in the UK. The main aim of this study is to inform the redesign of flexible energy-efficient housing units. The housing, designed for social tenants, was built by the Accent Group in 2005, using modern construction methods and sustainable materials, based on extensive research from the adaptable and "Grow Home" principles of Avi Friedman as well as open building implementation. The first pilot scheme was designed in collaboration with the Building Energy Analysis Unit at the University of Sheffield, together with the Goddard Wybor Practise, and was a successful housing development with respect to being environmentally friendly and a low-energy design scheme for the UK climate. This paper presents redesigning of flexible terraced housing units, and their performance evaluation, using a building simulation method as well as the passive-house planning package. The aim was to plan a row of terraced houses that can not only utilize a flexible design concept in floor planning layout, but also to reduce energy consumption with a passive design with particular attention paid to material selection. In addition, building simulation work has been carried out with the use of DesignBuilder software for both thermal and energy performance evaluation. The study examines the annual energy performance and comfort conditions in the designed house to be situated in the Northeast of England, UK. A terraced house unit design is considered a flexible home that can adjust to the needs of different tenants for the purpose of achieving a sustainable building under different aspects, such as low energy, low carbon, use of renewables, and low impact materials, with flexibility by design.

Reprinted from *Sustainability*. Cite as: Altan, H.; Gasperini, N.; Moshaver, S.; Frattari, A. Redesigning Terraced Social Housing in the UK for Flexibility Using Building Energy Simulation with Consideration of Passive Design. *Sustainability* **2015**, *7*, 5488-5507.

1. Introduction

Climate change is one of the biggest issues the world is facing today and the UK government is at the forefront to deal with this challenge. The domestic sector accounts for 27% of the UK's carbon emissions, making it necessary to pay attention to the way homes are designed and constructed [1]. Attention should also be given to making the existing housing stock energy efficient, as most of them will continue to be in existence for another 50 years, if not longer. According to the UK House of Commons Environment Committee, around 200,000 dwellings would need to be built each year in England, up to 2016, between 1991 and 2016, as there is a need for 4.4 million houses to be added to the housing stock [2]. According to the English Housing Survey [3], around 19% of the dwelling stock is flats; most of these are purpose built low-rise flats. The most common types of dwellings are terraced houses (29%), followed by semi-detached houses (26%). This was also stated in a report by the Office for National Statistics (ONS), which is the executive office of the UK Statistics Authority [4]. As the largest portion of homes in the country, it makes the existing terraced housing schemes not only a priority case for refurbishment and/or retrofit, but also offers a very good opportunity for redesign, to upgrade and/or improvement sustainability in the future.

The introduction of flexibility into the living space is the solution for users to adjust successfully to social behavioral changes. The Oxford dictionary describes the word "flexibility" as follows: "capable of bending easily without breaking" and "able to be easily modified to respond to altered circumstances or conditions" (Oxford 2015) [5].

Flexible architecture is described by Lelieveld [6] as follows:

"The term flexible architecture describes an architecture from which specific components can be changed in response to external stimuli, for example the users or environment. This change could be executed by the building system itself, transformed manually or could be any other ability to transform by an external force" [6] and [7] (p. 1).

The need for change is a universal phenomenon, and flexibility in the quality characteristics of a space accommodates spatial changes over a long period of time, rather than just thinking of the needs for present use of the space [8]. Furthermore, incorporating flexibility into the design can accommodate various functional demands within a limited space (economic benefit). In addition to its pragmatic benefits, flexible housing has good ecological potential, especially with regard to conserving energy and resources [9]. The refurbishment, obsolescence, and demolition of flexible designs require less material, energy, and labor, and, therefore, it is sustainable by design since there are less waste and lower costs (environmental benefit) associated with the process [10].

1.1. Flexible Housing

In the past few decades, architectural practice has faced a lot of challenges. Although our living conditions change in different contexts considering social, technological, economic, environmental, and political aspects, our housing types, responding to our patterns of living evolution, are still largely based on early 20th century models [11]. The speed of modernization, which started in the 18th century, is changing our lives, and the industrial revolution has made it so difficult for buildings to function without any flexibility, due to the changes in our lifestyle. Because of the mass migration of people from villages to cities, creating metropolises, social disorder, due to poor quality of living, has emerged and expanded because of new technology [12].

The architecture of flexibility can be considered one of the most important achievements of the Modern Movement [13]. This approach aimed to help dwellers change their living spaces with minimal costs; avoid expensive and destructive alterations; change their habits; and avoid being forced to move from their homes [14].

Schneider and Till in their paper "Flexible Housing; Opportunities and Limits" (2005), described flexible housing as "housing that is designed for choice at the design stage, both in terms of social use and construction, or designed for change over its lifetime" [15] (p. 157). There is a great need for studying flexibility of our living space with the potential of spatial changes over long periods of time rather than only thinking of the need of space for the present use. Architects and homebuilders need to surpass the phrase of "form follows function" in time and space.

1.2. Support, Infill and Open Building

N. John Habraken, a Dutch architect, educator, and theorist (1976) [16], employed the theory of the hierarchical principle of subsystems as one of the basic concepts of flexible housing. The essence of Habraken's theory is that, in housing, the role of community and the role of the individual are distinct. Habraken believed in reestablishing the dweller as an active participant in the total process of housing, and he developed a methodology for linking the processes of products and decisions to their technical implementation [16–19].

The division of "infill" and "support" was initially explored by Habraken, wherein, he explained support as the permanent part of a building provided to the occupant by the builder or the architect commissioned; while infill includes the interior partitions, a kitchen, and bathrooms as defined by the occupant. The systems and parts associated with the infill level tend to change in cycles of 10 to 20 [17,19]. These transformations may be occasioned by the occupants' changing requirements or preferences; by the cyclical need for technical upgrades; or by changes in the base building (support) [19]. The open building (OB) implementation approach is the continuation of Habraken's approach, elaborating more on practical aspects, design, and construction [10].

In this paper, a case study of a pilot terraced energy-efficient social housing unit (Figure 1) has been chosen, with the main aim being to inform the redesign of flexible energy-efficient housing units. The housing was designed for social tenants, and built by the Accent Group in 2005 using modern methods of construction (MMC), and sustainable materials based on extensive research from the "adaptable" and "grow home" principles of Avi Friedman [20]. The first pilot scheme was designed in collaboration with the Building Energy Analysis Unit (BEAU) at the University of Sheffield [21] together with architects at the Goddard Wybor Practise (GWP) Ltd. The project was a successful housing development. Being both environmentally friendly and a low energy design scheme for the UK climate.

This paper presents the redesign of terraced housing units [22] and their energy performance evaluation, using building simulation together with support from a passive-house planning package. The aim is to plan a row of terraced houses that can utilize a flexible design concept as well as building performance evaluation for optimizing energy demand and planning of passive design. Moreover, the terraced house design is considered a flexible home that can be adjusted the tenants' different needs, with the purpose of achieving sustainability with respect to different aspects, such as low energy, low carbon, use of renewables and low impact materials, with flexibility by design. This project focuses on the flexible design of infill (interior of unit). Furthermore, it compares the performance evaluation of the flexible floor plan layout and the traditional existing layout.

Figure 1. A row of terrace houses—energy efficient social housing in northeast England, UK.

2. Approach, Methodology and Analysis

2.1. Early Design Stage

The case study chosen for the studies in this paper goes back to the "Accent Home" concept, a terraced house design for social housing, developed and situated in Leeds/Bradford region in the northeast of England. The building design was completed by one of the UK's largest social housing developers at the time to achieve low cost and low energy designs for the UK social housing sector.

The aim of this study was to design a new sustainable building prototype situated in the same area as the Accent Home; beginning from the early stages of architectural design, together with a vision to develop future housing with flexibility in mind, to meet the needs of different tenants or the needs of a single family in the future. Going back to the drawing board, all dimensions of each apartment have been considered. On an overall plan of 290 m^2, there are four apartments (*i.e.*, units); each one a rectangular plan of 12 m (north–south direction) × 6 m (east–west direction). The inclination of 19° to the east is to maximize the sunshine in the morning as opposed to the afternoon. Moreover, solar analyses have been conducted because of the importance of sun and its directions from different periods in a typical year as well as the size and length of shading devices to avoid excessive solar gains during the summer period (Figure 2a,b).

After studying the history of terraced houses in England (UK) and considering design flexibility, the design started to take a different configuration that could arrive at the goal of flexibility, while at the same time be energy efficient, before finally arriving at the chosen layout [15,22].

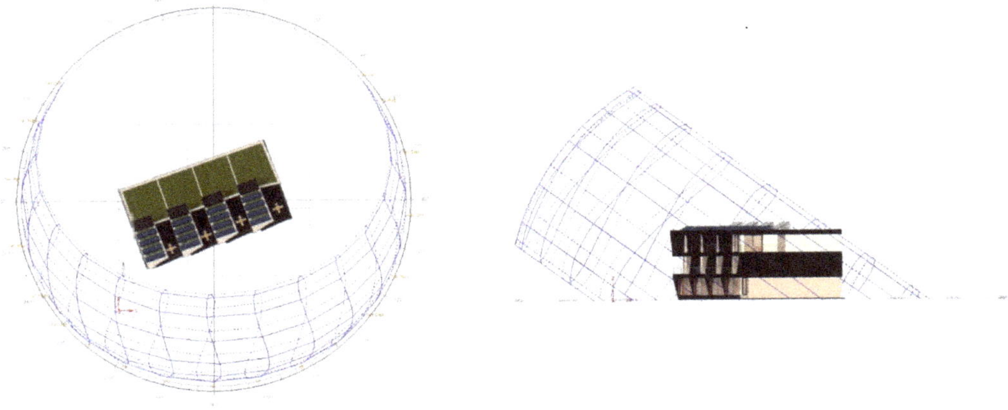

(a) (b)

Figure 2. Annual sun path: top view (**a**); east front (**b**).

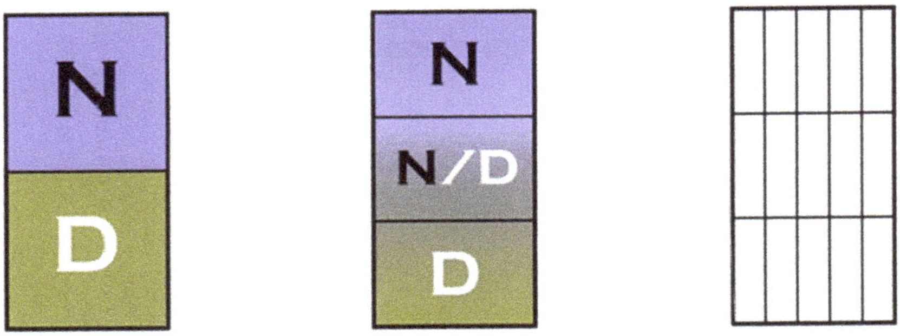

(a) (b) (c)

Figure 3. Layout and configuration (N: Night, D: Day).

As mentioned earlier, the main facade of the building is oriented 19° to the east and the extended plan is in the shape of a rectangle (12 m × 6 m). First, a decision has been given to divide the plan into two different square areas: Day (D) and Night (N) (Figure 3a,b). The Day area is suitable for daily activities, due to natural light and solar gains coming from the south direction. On the other hand, the Night area is situated at the north side of the house, which is perfect for nightly activities that do not require too much light (Figure 3a). The first division of the apartment into two different areas was too straight for future movement of the building interior space. Dividing the length into three different areas allowed defining a third space in between, so that this could be used for both daily and nightly activities. This new space can be called "Flexible Space" or "Margin", also named by Habraken, "adaptability" (Figure 4b) [16]. The short side of the apartment needed a division that could take into account all the different aspects of the house elements, such as stairs, doors, furniture, appliances, *etc.* A 120 cm module (*i.e.*, a full module) is used to fit every element into a regular grid. For example, a half module (60 cm) is enough for a kitchen space with furniture and appliances; a 2/3 module (80 cm) is enough space for doors and a single bed; and a full module (120 cm) is enough space for stairs and sliding doors (Figure 3c).

The positioning of the staircase and washroom is an important decision that would define all the following phases of a flexible design. These are, in effect, two inflexible elements that once placed cannot be easily moved. For this reason, decision of situating the staircase in the north side of the building, instead of the washroom, needed accurate analysis. The washroom in the back might have an external window that would also allow natural ventilation. On the other hand, the stairs in that position creates an independent nucleus and a new entry to the building, and to each independent level of the building. This was considered the most flexible solution (Figure 4).

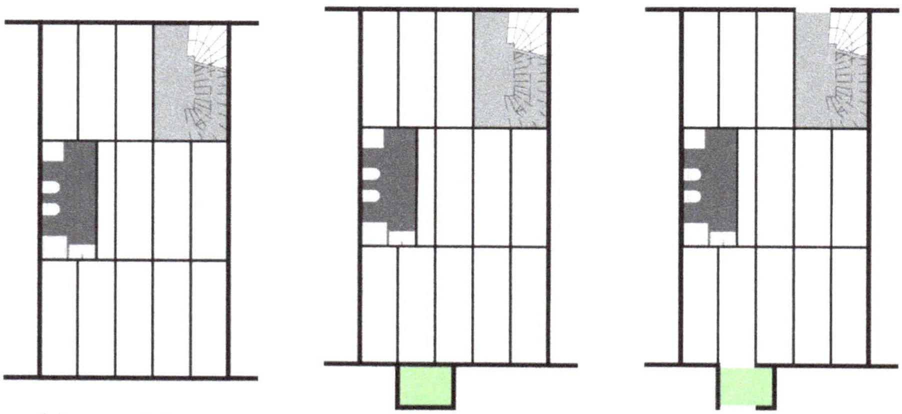

(a) ground floor (b) first floor (c) second floor

Figure 4. Positioning of continues staircase and toilettes on all three floor plans.

Furthermore, the stairwell can be used for natural ventilation inside the building and can create a natural solar chimney (Figure 5a). As a consequence, the bathroom found the right arrangements at the opposite side of the central area. In this particular situation, an inflexible space such as the bathroom, helped to create a distinct division of the apartment into three different areas, and at the same time, the possibility of considering all the remaining spaces as unique zones (Figure 4b).

The positioning of the main entry is another important problem to be solved in a flexible design and, moreover, in a terraced home. First, the main entry has a double function, *i.e.*, it is entranceway from the main street and is also a space that can be used as a temporary deposit, and a small greenhouse in the wintertime. Moreover, the off-center position permits the creation of two spaces, for big and small windows, a solution that allows a division of the interior space in the future. The distance from the partition wall permits solar rays coming from the east to reach the big window of the preceding apartment (Figure 4c).

Due to the main entry and the staircase disposition, the apartment is provided with two independent entry doors. The first one on the south facade and connected directly with the main street, and the second one on the north side that can be used as a service door or an independent door to arrive at the apartments of the first and second floors. The internal space resulting from this division is a long and dynamic space of 48 m^2 that can meet the needs of different tenants (Figure 5a).

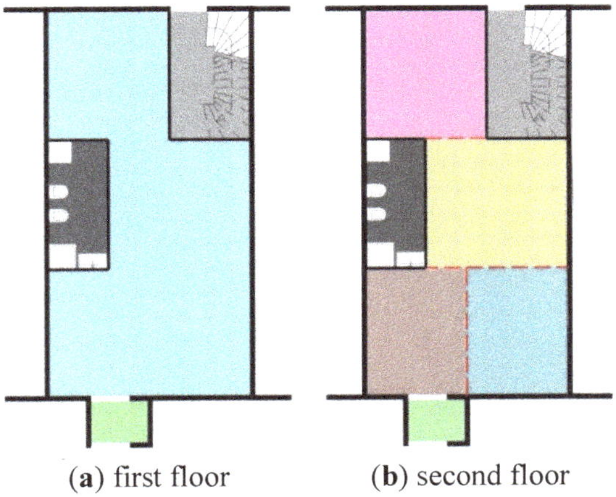

Figure 5. Spaces with possible divisions on upper floors.

Figure 5 shows a top view of the space, with possible divisions. Tenants can convert different zones using sliding walls, *i.e.*, sliding walls for daily changes or through more or less flexible walls (Figure 5b).

Figure 6 is the social housing development pilot project as a row of terraced houses, which has been redesigned with flexibility, to meet the needs of different tenants including, single-family users in the future.

Figure 6. Redesign of a social housing development.

The division of infill and support was initially explored as one of the main principles of flexible housing. This division support is the permanent part of the building provided to the occupant by the builder or the architect commissioned, while infill includes the interior partitions, for example a kitchen and bathrooms, as defined by the occupant. The systems and parts associated with the infill level tend to change in cycles of 10 to 20 years [16], whereas the support has a 100-year lifespan.

These transformations may be occasioned by the occupants' changing requirements or preferences, by the cyclical need for technical upgrades, or by changes in the base building (support). In this project, the clear span of floor plan is proposed as the support, whereas the infill (interior partition) is being customized gradually.

Seeing the possible divisions (*i.e.*, configuration layouts) (Figure 5a,b) and the longitudinal section (Figure 7), helps to understand the way in which the housing units can grow over their lifespan. Starting from a configuration with one unit, *i.e.*, first and second floor units (Figure 5a,b), it is possible to divide the same housing unit into two or even three separate units (Figure 8, left to right).

Figure 7. A longitudinal section with growing units.

The internal space resulting from the division is a long and dynamic space of 48 m^2 that can meet the needs of different tenants. In addition, Tenants can convert different zones through sliding walls, or through more or less flexible walls. In the following, the full layouts of the housing units' floor plans are presented (Figure 8).

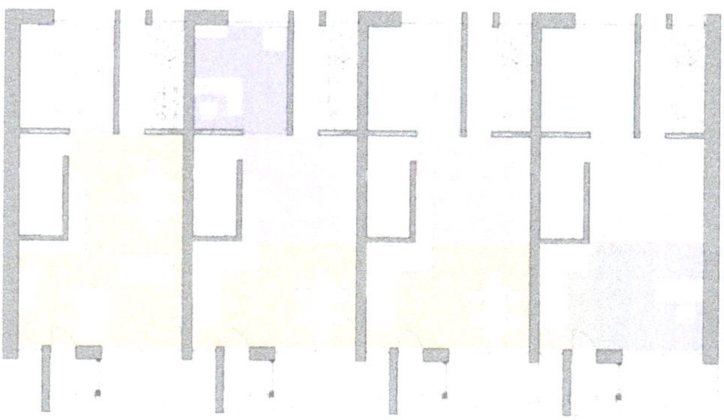

Ground Floor

Figure 8. *Cont.*

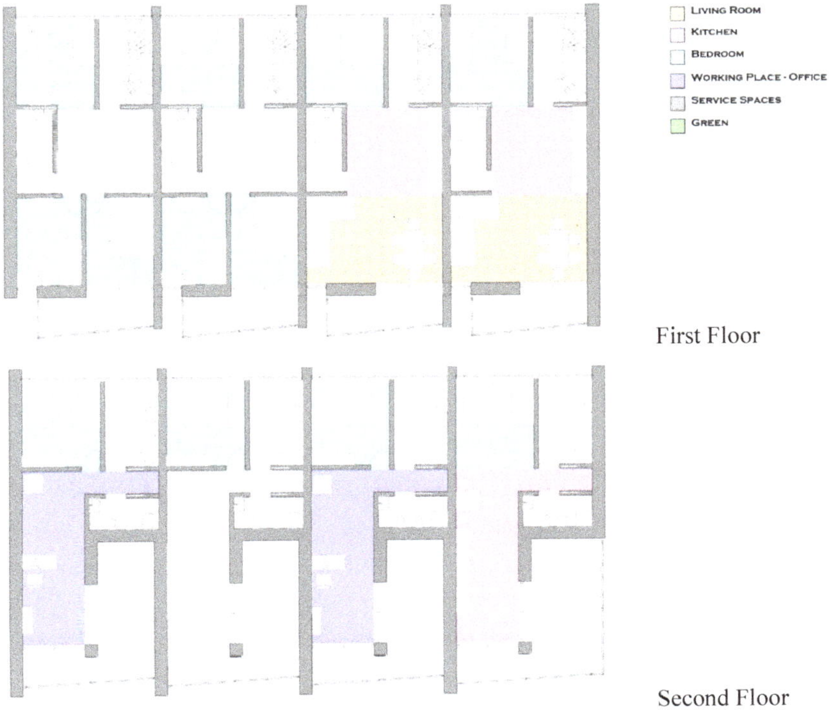

Figure 8. Floor plans of the housing units.

2.2. Passive House Planning

Energy efficiency in housing is an important factor that has to be considered in order to achieve the aim of sustainable housing. Thus, in recent years, the applications of renewable energy in buildings has rapidly increased; sometimes supported by national subsidies. Nevertheless, the high costs of energy systems and their complex integration, coupled with expensive upkeep, discourages many people from investing in such systems. It is important to design flexible houses in order to advance the use of new technologies; however, it is equally necessary to reduce the use of energy systems by maximizing the advantages of the sun's energy, location and materials. This is a passive method that allows comfort with few energy systems.

Sustainable building, through the Passivhaus standard [23], can be achieved at any location in the world. Passive houses are very well insulated and draught-proofed buildings whose annual space heat demand is so low that a conventional heating system can be omitted. The small amount of heat still required can be delivered to individual rooms by heating the air supplied through the ventilation system. This occurs when the space heating energy demand is not more than that of 15 kWh/m^2 per annum [23].

Super insulation, isolated thermal bridges and airtightness are three important principles of the Passivhaus standard that have to be considered during every design stage. In this flexible and passive house design study, there are low *U*-values [24] (*i.e.*, thermal conductivity or heat transfer coefficient—a measure of heat loss in a building element such as a wall, floor or roof); windows with u-values of 0.77 W/m²K, external walls with u-values of 0.11 W/m²K, a roof with a u-value of 0.11 W/m²K, and a ground floor with a u-value of 0.12 W/m²K, which can noticeably reduce the heat transfers.

In the case study home, a combination of the planning principles from the "plotter pen" [23] and a cross-laminated timber structure to avoid thermal bridges through the building boundary has been considered (see Figure 9).

The following section (Figure 9) shows the planning principle from the plotter pen. This principle is used in the design of a passive house, and in particular in sections and in floor plans. The width of a plotter pen corresponds to a heat transmission resistance of R = 6 m²K/W. In this case, the following equation is used:

$$R = \frac{s}{\lambda} \tag{1}$$

where R is the thermal resistance, λ is the thermal conductivity (u-value), s is the thickness of an insulation material with thermal conductivity value of 0.043 W/(mK) (wood soft fiber). In the study, a scale width of 25.8 cm is approximated at 25 cm (0.25 m). If the outer shell (red insulation layer in Figure 10) manages to wrap the envelope surrounding the whole building with full width and without any interruption, the project can avoid thermal bridges (except for more detailed analysis).

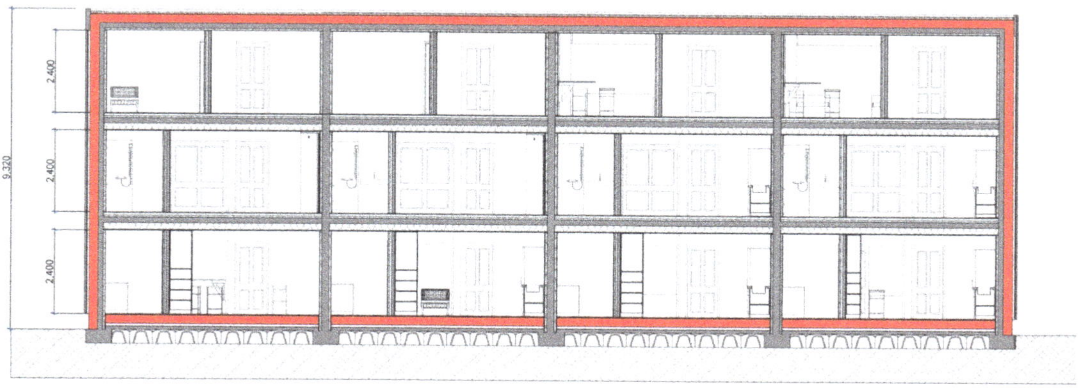

Figure 9. Planning principle from the "plotter pen" in east–west section.

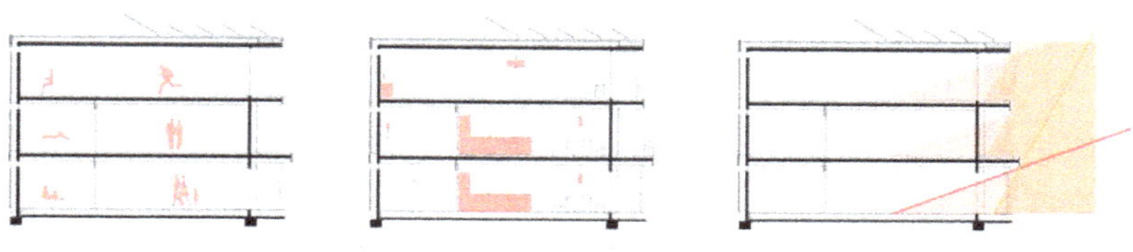

(a) (b) (c)

Figure 10. Internal gain model (**a**–**c**).

The Passive House Planning Package (PHPP) has been used as a design tool in the process of redesign. The package provides the architect and the engineer with instruments that are necessary for the design of a well-performing passive house [23]. Thus, the definition of PHPP needs some more clarification. In PHPP calculations, not only the heat losses, but also the heat gains and the thermal inertia can be determined. Through the PHPP worksheets, it is possible to define all heat gained from the sun, appliances, domestic hot water (DHW) distribution, and people as the users of the building. The latter internal gain is not a constant value in a flexible housing and has to be considered under different kinds of occupancy.

The first aspect noticed while working with the PHPP worksheet is the straight-forward organization of Microsoft Excel based spreadsheet. It is easy to insert all the different values for a building that is regular in plan or in elevation, however it is not the same way for a building that has an irregular shape or, such as in this case, a prototype that needs little adjustments in the architectural design in order to find the best shape for good indoor comfort. Another aspect that did not help in the design of a flexible home is the calculation of PHPP on a treated floor area. One of the main objectives of this project was to understand the difference in terms of the indoor comfort between the different kinds of configurations that a flexible design could allow; however, with the PHPP, only the comfort for the overall treated area of the whole building can be calculated. This aspect, for example, has importance for studying the overheating hours. The PHPP verification worksheet gives the frequency of overheating (Figure 11, central column, 7th row), which is the hour percentage per year in which the temperature exceeds 25 °C. Despite the fact that this result gives a good approximation of the average temperature, it does not give any information about the temperature distribution because there is no distinction between spaces in the north or south direction, or between the utilization of every room.

Energy performance evaluation is usually beneficial at the early stages of design; however, in this particular case, in redesigning of a social housing development, it is still an important factor that has to be considered in order to make the necessary provisions for achieving the best practice standards, *i.e.*, low energy and low carbon [25,26] for the future of sustainable housing. Hence, in recent years, applications of low and zero carbon technologies, and the adoption of renewable energy sources in domestic buildings has rapidly increased. On the one hand, it is equally important to design flexible houses in order to create provisions for the needs of future lifestyles, while also reducing the use of energy demand by maximizing the advantages of passive design and the use of low impact materials. This approach is essential nowadays and the benefit of passive design could allow indoor comfort with few renewable technologies. Passive design is not a supplement to architectural design, but it is rather part of the design process itself through an integrated approach to architectural and sustainable building design [27–29].

Passive House Verification

Building:	Flex PH Housing	
Location and Climate:		Leed Monthly Mean
Street:	Barnby Pl	
Postcode/City:	BD2 4RB	
Country:	England	
Building Type:	Terraced House	

Year of Construction:	2013				
Number of Dwelling Units:	1		Interior Temperature:	20,0	°C
Enclosed Volume V_e:	714,0	m^3	Internal Heat Gains:	2,1	W/m²
Number of Occupants:	4,2				

Specific Demands with Reference to the Treated Floor Area

	Applied:	Annual Method		PH Certificate:	Fulfilled?
Treated Floor Area:	147,5	m²			
Specific Space Heat Demand:	**14**	**kWh/(m²a)**		**15 kWh/(m²a)**	**Yes**
Pressurization Test Result:	**0,2**	**h^{-1}**		0,6 h^{-1}	**Yes**
Specific Primary Energy Demand (DHW, Heating, Cooling, Auxiliary and Household Electricity):	**110**	**kWh/(m²a)**		120 kWh/(m²a)	**Yes**
Specific Primary Energy Demand (DHW, Heating and Auxiliary Electricity):	68	kWh/(m²a)			
Specific Primary Energy Demand Energy Conservation by Solar Electricity:	34	kWh/(m²a)			
Heating Load:	9	W/m²			
Frequency of Overheating:	0	%	over 25 °C		
Specific Useful Cooling Energy Demand:		kWh/(m²a)		15 kWh/(m²a)	
Cooling Load:	2	W/m²			

Figure 11. Passive House Planning Package (PHPP) verification worksheet.

2.3. Building Simulation

In the study, DesignBuilder software has been used for conducting building simulation work [30]. The software package has integration with EnergyPlus, the US Department of Energy's (DOE) third generation dynamic building energy simulation engine for modeling building, heating, cooling, lighting, ventilating and other energy flows. This integration within DesignBuilder allow the user to carry out complete simulations without having to leave the interface, which is an excellent feature for practicing architects. The following table shows the necessary input information for building simulation analysis (Table 1).

Table 1. Building simulation input data for Heating, Ventilation, and Air Conditioning (HVAC), Domestic Hot Water (DHW), activity and construction.

Compact HVAC	
HVAC systems is defined parametrically and modeled within EnergyPlus using Compact HVAC descriptions with a CAV (constant air volume)	
Natural Ventilation	
Natural Ventilation and infiltration air flow rates is calculated based on opening and crack, sizes, buoyancy, wind pressure and the activity schedules	
Mechanical Ventilation	
Mechanical ventilation utilized in the flexible housing design has an ac/h (air change per hour) rate of 0.4, as required by the Passivhaus standard with an outside air definition method set for zone	
Fans	
Night cycle control	Cycle on control zone
Fan placement	Blow through
Part-load power coefficients	Variable speed motor
Fan type	Intake
Pressure rise (pa)	1000.0
Total efficiency (%)	85.0
Fan motor in air (%)	100.0
Outside air definition method	Minimum fresh air (Per area)
Outside air mixing	Recirculation
Outside air control minimum flow type	Proportional
Heat recovery type	Sensible
Sensible Heat Recovery Effectiveness	0.800
Heating set-point temperature	15.00
Domestic Hot Water (DHW)	
Type	Dedicated DHW boiler
DHW CoP	0.85
Fuel	Biomass
Water Temperatures	
Delivery temperature (°C)	65.00
Mains supply temperature (°C)	10.00
Activity	
- Compact schedule has been used for occupancy, metabolic activity, openings of windows and doors, lighting, and the schedules are based on data published in the UK's National Calculation Methodology (NCM) * - * The NCM for the EU's Energy Performance of Buildings Directive (EPBD) is defined by the Department for Communities and Local Government (DCLG) [31].	
Construction	
DesignBuilder uses construction components to model the conduction of heat through walls, roofs, ground and other opaque parts of the building envelope. Using the construction data, the physical properties of each element have been defined for the building (e.g., external wall, party wall, interior wall, roof, floors and ground floor). The same is true for windows and doors that can be selected from a well-provided library.	

3. Results and Discussions

In this section, the analysis based on simulation studies has been presented. For the simulations, the following table shows several details of the selected housing unit (Table 2). As part of the analysis, a series of simulations has been carried out using DesignBuilder software considering the above input data (Table 1). One of the concerns was the overheating potential due to the type of construction used and in the case of Passivhaus standard adopted [23,32], *i.e.*, super insulation, high levels of airtightness, and large levels of glazed south facade in the house design. Therefore, the study evaluated the internal comfort during one of the warmest weeks of the year in order to avoid likely overheating during summer periods.

Table 2. Selected housing unit information.

Dimensions	Meters (m)
Width	6.4
Depth	9.0
Floor Height	2.4
Type of Unit	Mid-terraced House
Number of Floors	Ground + 2
Number of Bedrooms	4
Floor Area	170 m^2
Climate Data	Leeds/Bradford (UK)

The following graphs are the results of an hourly internal comfort and temperature distributions for the living room on the ground floor from 6 to 13 July 2013 (Figures 12 and 13).

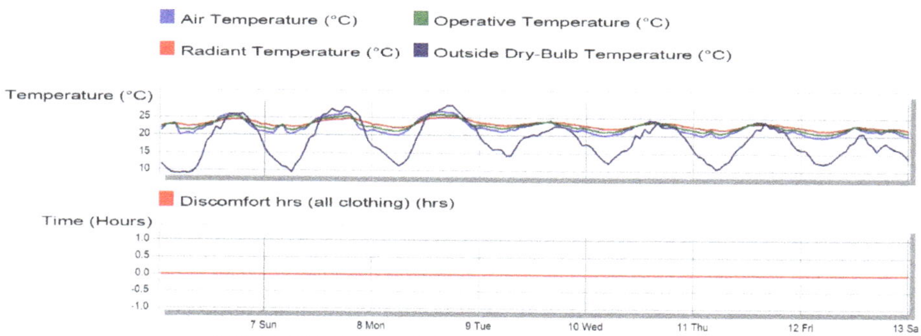

Air Temperature (°C)	25.45	20.31	21.97	25.80	21.66	22.89	21.81	20.72	24.03	19.75
Radian Temperature (°C)	24.13	22.35	24.18	24.48	23.21	23.80	22.42	22.53	23.53	21.51
Operative Temperature (°C)	24.79	21.33	23.07	25.14	22.44	23.34	22.12	21.63	23.78	20.63
Outside Dry-Bulb Temperature (°C)	23.80	9.20	25.60	24.90	15.90	22.60	15.90	15.90	21.90	12.40

Figure 12. *Cont.*

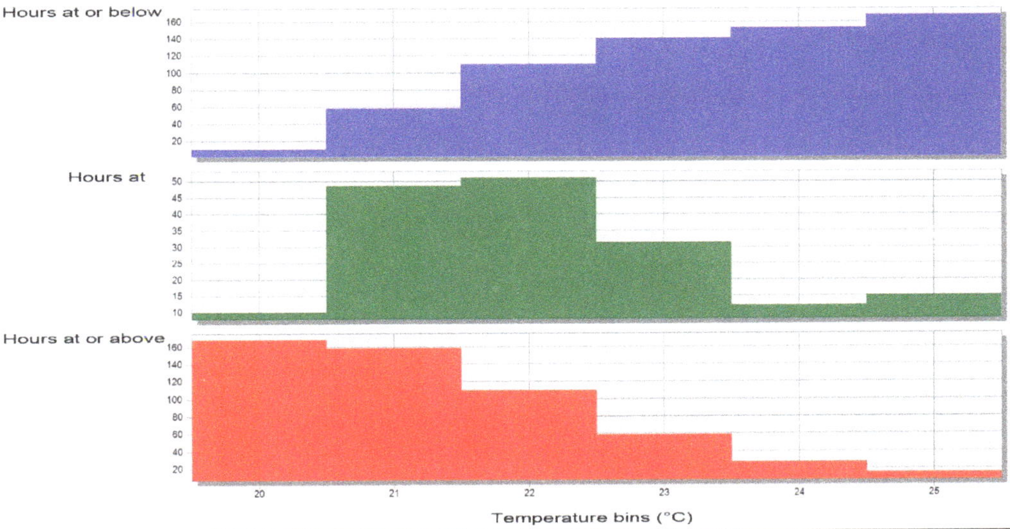

	20	21	22	23	24	25
Hours at or below	10.0	58.5	109.5	141.0	153.0	168.0
Hours at	10.0	48.5	51.0	31.5	12.0	15.0
Hours at or above	168.0	158.0	109.5	58.5	27.0	15.0

Figure 12. Typical summer week showing temperature and discomfort hours.

Again, same evaluation has been carried out for a typical winter week and the results have shown that the average temperature profile stays constant around 20 °C due to heat recovery, insulation and solar gains (Figure 13). In a passive house design with the Passivhaus standard, there is no use of a space heating system, but there is a system that contains mechanical ventilation with heat recovery (MVHR) (Table 1).

The following graph presents the total fuel breakdown listing different energy consumption for the redesigned flexible housing unit (Figure 14).

As the housing unit chosen has a floor area of 170 m^2, to obtain the annual energy consumption and the associated carbon dioxide (CO_2) emissions, further calculations have been made to present the necessary values (Table 3).

In order to demonstrate the benefits of simulation study findings for the case of redesign, it is important to compare against the measured consumption figures using the actual data recorded during post occupancy monitoring. The main reason for this is to cross-check the results with the existing social housing units (*i.e.*, to compare with a real non-Passivhaus case where the units were designed to be energy efficient) and for the validity of the simulation studies. In this case, the following table presents the average measured energy consumption and the associated carbon dioxide emissions obtained from on-site monitoring between 2007 and 2009 (Table 4) [33]. As the existing housing unit floor area is 128 m^2, the calculations are based on this value.

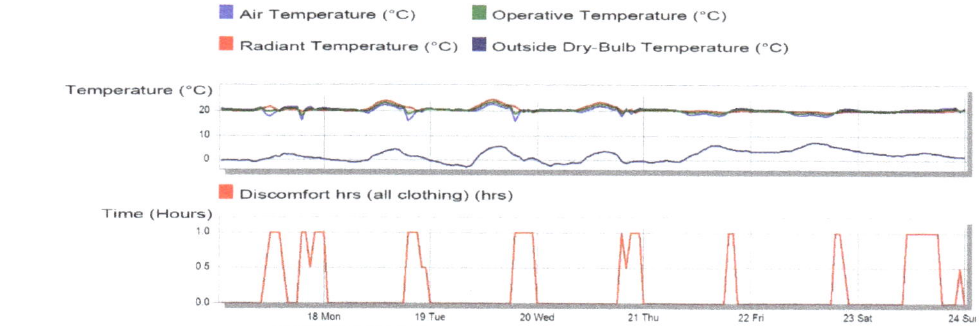

Air Temperature (°C)	21.55	20.68	20.49	21.97	20.81	21.03	19.80	21.01	18.74	20.50
Radian Temperature (°C)	20.59	20.14	19.80	23.79	20.50	21.91	20.48	20.56	20.31	20.08
Operative Temperature (°C)	21.07	20.41	20.14	22.88	20.65	21.47	20.14	20.78	19.53	20.29
Outside Dry-Bulb Temperature (°C)	2.40	−1.10	1.30	3.40	−2.40	2.20	0.30	3.60	7.30	2.70

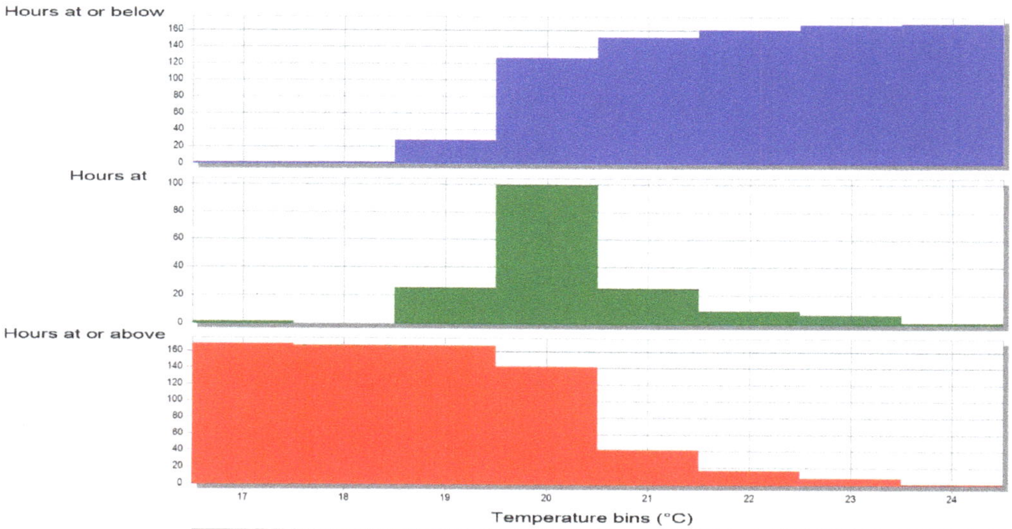

Hours at or below	1.5	1.5	27.0	126.5	151.5	160.5	167.0	168.0
Hours at	1.5	0.0	25.5	99.5	25.0	9.0	6.5	1.0
Hours at or above	168.0	166.5	166.5	141.0	41.5	16.5	7.5	1.0

Figure 13. Typical winter week showing temperature and discomfort hours.

As can be seen from Tables 3 and 4, the simulated housing unit, *i.e.*, the redesigned housing unit, has shown much better performance in terms of both the energy demand required and the carbon dioxide emitted. This, however, should be validated through real-time measurements in the redesigned units, particularly for checking against the accuracy of simulation studies.

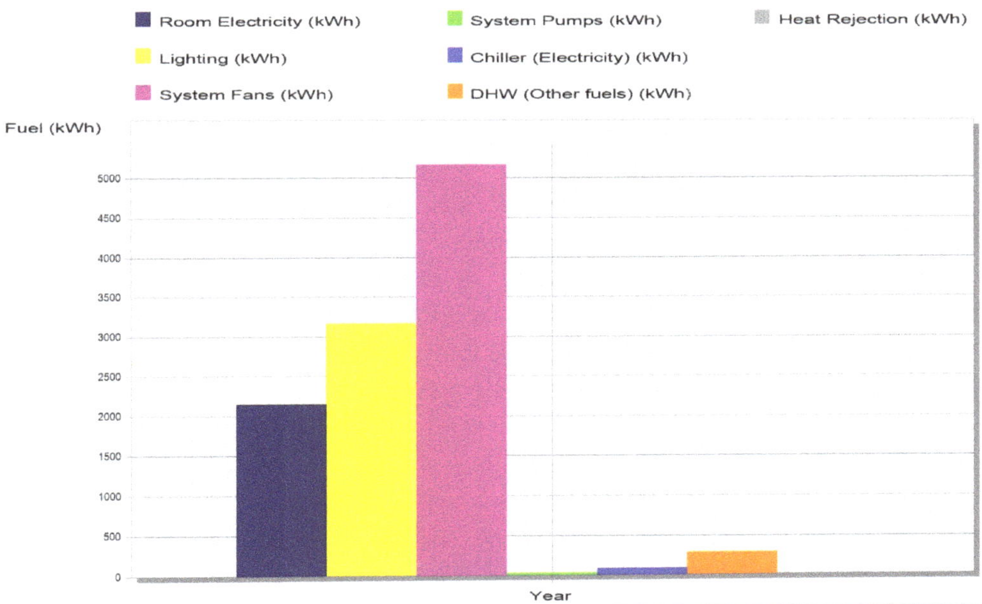

Room Electricity (kWh)	2148.61
Lighting (kWh)	3161.46
System Fans (kWh)	5157.22
System Pumps (kWh)	31.13
Chiller (electricity) (kWh)	84.65
DHW (Other fuels) (kWh)	296.06
Heat Rejection (kWh)	1.07

Figure 14. Fuel breakdown.

Table 3. Annual energy consumption and carbon emissions.

Total Energy Demand	Carbon Dioxide (CO_2) Emission
10,880 kWh/year	7493 kg/year
64 kWh/m^2	**44** kg/m^2

Table 4. Measured energy consumption and carbon dioxide emissions.

Total Energy Demand	Carbon Dioxide (CO_2) Emission
13,358 kWh/year *	6833 kg/year *
~**104** kWh/m^2	~**244** kg/m^2

* The measured data is based on an average usage during years 2007 to 2009 [33]. Note: Floor area for the same unit is 128 m^2.

4. Conclusions

This paper attempts to describe an approach taken for different design phases to propose a new sustainable building, starting from the early stages of design to the building simulation component for performance evaluation. Collaboration between architects and engineers (*i.e.*, aiding the design of a sustainable building; in this case a sustainable social housing unit) helped to find indoor comfort solutions by combining both structural and architectural features with integration of energy systems, where building simulation task played an important role.

PHPP has been very useful to acquire specific energy and heat demand analyses, and to verify whether the building achieves the Passivhaus standard. On the other hand, DesignBuilder can simulate indoor comfort conditions in greater detail, which makes it easier for the designer to understand the Passivhaus planning stages and how best to manage the results obtained through changing parameters and finding new solutions to fulfill all the requirements set to meet such a standard. Despite two software packages being used for various analyses in the study, both PHPP and DesignBuilder have many features, and provided similar results when compared to each other; thus, the use of both platforms helped to conduct a complete environmental analysis of the building designs, allowing cross-check of results.

The redesign of a terraced home has been a worthy challenge in terms of adopting flexibility in the building form and layout, as well as achieving the best practice energy efficiency standards. In this way, it was also useful to utilize building simulation and performance evaluation as means to help aid the building design, considering various important factors, such as the Passivhaus standard, concerns of overheating, and energy efficiency. Moreover, in real life situations, the occupant (*i.e.*, the user) could still make a big difference in the way the energy is being used at home, even though the design could be a high performance case. However, it is still vital to get the design specification right and tested throughout the design process, which has been demonstrated in this study. Therefore, without undermining the energy conscious user/behavior to save energy at home, further studies should be undertaken and also related to the people variable, *i.e.*, their behavior with attention paid to energy and environment, as a follow-up task.

Last but not least, this is a reminder for all practicing architects and design professionals to utilize the available building simulation platforms to aid their designs, or redesigns, at the early stages of design and throughout their design process. It is also equally important to aid the design with planning of passive design with the use of PHPP. Once more, integrated building design is not a myth, but a way of today's building design practice, and all professionals should work closely in full collaboration and integration, such as architecture and engineering, to help solving challenges of building science.

Acknowledgments

The authors would like to thank the Accent Group Ltd. and their architects, the Goddard Wybor Practice (GWP) Ltd., and the Building Energy Analysis Unit (BEAU) at the University of Sheffield for providing the grounds and information to carry out this study. The project was supported by the EU's Lifelong Learning Programme (LLP).

Author Contributions

Hasim Altan designed the research. Nicola Gasperini performed the research and analyzed the data under supervision of Hasim Altan. Sam Moshaver provided academic support in flexible housing and Antonio Frattari provided academic advice throughout the research. All authors equally contributed in the writing of this paper. All authors read and approved the final manuscript.

Conflicts of Interest

The authors declare no conflict of interest.

References

1. Department of Energy and Climate Change (DECC). Digest of United Kingdom Energy Statistics. Available online: https://www.gov.uk/government/publications/energy-chapter-1-digest-of-united-kingdom-energy-statistics-dukes (accessed on 10 March 2014).
2. Golland, A.; Blake, R. *Housing Development: Theory, Process and Practice (Housing, Planning and Design Series)*; Routledge: London, UK, 2004.
3. Department for Communities and Local Government (DCLG). English Housing Survey: Housing Stock Report 2009. Available online: https://www.gov.uk/government/uploads/system/uploads/attachment_data/file/6725/1937212.pdf (accessed on 27 March 2015).
4. Office for National Statistics (ONS). Housing: Social Trends 41. Available online: http://www.ons.gov.uk/ (accessed on 15 April 2015).
5. Oxford. Oxford Dictionaries. Available online: http://www.oxforddictionaries.com/ (accessed on 24 March 2015).
6. Lelieveld, C.M.J.L.; Voorbij, A.I.M.; Poelman, W.A. Adaptable Architecture. In *Building Stock Activation*; Kitsutaka, Y., Ed.; TAIHEI Printing Co.: Tokyo, Japan, 2007; pp. 245–252.
7. Cuperus, Y. An introduction to Open Building. In Proceedings of the 9th Annual Conference of the International Group for Lean Construction (IGLC 2001), Singapore, 6–8 August 2001.
8. Altaş, N.E.; Özsoy, A. Spatial Adaptability and Flexibility as Parameters of User Satisfaction for Quality Housing. *Build. Environ.* **1998**, *33*, 315–323.
9. Narahara, T. Designing for Constant Change: An Adaptable Growth Model for Architecture. *Int. J. Archit. Comput.* **2010**, *8*, 29–40.
10. Kendall, S.; Teicher, J. *Residential Open Buildings*; E & FN Spon: New York, NY, USA, 2010.
11. Forster, W. *Housing in the 20th and 21st Centuries*, Bilingual ed.; Prestel Publishing: New York, NY, USA, 2006.
12. Leupen, B.; Heijine, R.; Zwol, J.V. *Time Based Architecture*; 010 Publishers: Rotterdam, The Netherlands, 2004.
13. Henket, H.J.; Heynen, H. *Back from Utopia: The Challenge of the Modern Movement*; 010 Publishers: Rotterdam, The Netherlands, 2002.
14. Lans, W.; Hofland, C. Flexibility, How to Accommodate Unknown Future Housing Requirements. In Proceedings of the 30th IAHS World Congress on Housing, Transforming Housing Environments through Design, Pretoria, South Africa, 27–30 September 2005.
15. Schneider, T.; Till, J. Flexible Housing: Opportunities and Limits. *Archit. Res. Q.* **2005**, *9*, 157–166.

16. Habraken, N.J.; Boekholt, J.T.; Dinjens, P.J.M.; Thijssen, A.P. *Variations: The Systematic Design of Supports*; MIT Press: Cambridge, MA, USA, 1976.
17. Cheng, M.-Y.; Lien, L.-C.; Tsai, M.; Chen, W.-N. Open-building Maintenance Management Using RFID Technology. In Proceedings of the 24th International Symposium on Automation and Robotics in Construction (ISARC 2007), Kochi, India, 19–21 September 2007.
18. Dluhosch, E. Flexibility/Variability and Programming. In *Industrialization Forum*; University of California: Berkeley, CA, USA, 1974; Volume 5, No 5.
19. Moshaver, S.; Altan, H. A Knowledge Model to Implement Home Working in Multi-tenant Housing. In Proceedings of the 15th International Union of Architects (UIA) World Congress, (DURBAN 2014), Architecture Otherwhere, Resilience—Ecology—Values, Durban, South Africa, 3–7 August 2014; p. 63.
20. Friedman, A. *The Adaptable Home: Designing Homes for Change*; McGraw-Hill: New York, NY, USA, 2002.
21. Altan, H. Sustainable Planning and Design: The Role of Environmental Analysis at the Initial Design Stages. In Proceedings of the 19th International Association for People-Environment Studies (IAPS) Conference, Alexandria, Egypt, 11–16 September 2006.
22. Muthesius, S. *The English Terraced House*; Yale University Press: New Haven, CT, USA, 1982.
23. Passivhaus Institut (PHI). *Passive House Planning Package (PHPP) 2007: User Guide*; PHI: Darmstadt, Germany, 2010.
24. Greenteg. Heat Flux Sensors and U-Value Kit for Building Physics. Available online: http://www.greenteg.com/heat-flux-sensor/u-value-measurement/ (accessed on 22 April 2015).
25. Energy Saving Trust (EST). *Energy Efficiency Best Practice in Housing: Energy Efficient Refurbishment of Existing Housing*; EST: London, UK, 2004.
26. Energy Saving Trust (EST). *Domestic Low and Zero Carbon Technologies: Technical and Practical Integration in Housing*; EST: London, UK, 2010.
27. Brophy, V.; Lewis, J.O. *A Green Vitruvius: Principles and Practice of Sustainable Architectural Design*, 2nd ed.; Earthscan: London, UK, 2011.
28. Keeler, M.; Burke, B. *Fundamentals of Integrated Design for Sustainable Building*, 1st ed.; US Green Building Council: NJ, USA, 2009.
29. Yudelson, J. *Green Building through Integrated Design*; McGraw-Hill Professional: New York, NY, USA, 2009.
30. DesignBuilder. DesignBuilder Software Energy Simulation Package. Available online: http://www.designbuilder.co.uk/ (accessed on 2 August 2013).
31. BRE. National Calculation Method (NCM). Available online: http://www.ncm.bre.co.uk/ (accessed on 8 August 2011).
32. Uffelen, C.V. *Passive Houses: Energy Efficient Homes*; Braun Publishing AG: Salenstein, Switzerland, 2012.
33. Altan, H. *Accent Homes Project: Post-Occupancy Monitoring Internal Report*; Sheffield University: Sheffield, UK, 2010.

Environmental Impacts and Embodied Energy of Construction Methods and Materials in Low-Income Tropical Housing

Arman Hashemi, Heather Cruickshank and Ali Cheshmehzangi

Abstract: This paper evaluates the current conditions of Ugandan low-income tropical housing with a focus on construction methods and materials in order to identify the key areas for improvement. Literature review, site visits and photographic surveys are carried out to collect relevant information on prevailing construction methods/materials and on their environmental impacts in rural areas. Low quality, high waste, and energy intensive production methods, as well as excessive soil extraction and deforestation, are identified as the main environmental damage of the current construction methods and materials. The embodied energy is highlighted as the key area which should be addressed to reduce the CO_2 emissions of low-income tropical housing. The results indicate that the embodied energy of fired bricks in Uganda is up to 5.7 times more than general clay bricks. Concrete walling is identified as a much more environmentally friendly construction method compared to brick walling in East African countries. Improving fuel efficiency and moulding systems, increasing access to renewable energy sources, raising public awareness, educating local manufacturers and artisans, and gradual long-term introduction of innovative construction methods and materials which are adapted to local needs and conditions are some of the recommended actions to improve the current conditions.

Reprinted from *Sustainability*. Cite as: Hashemi, A.; Cruickshank, H.; Cheshmehzangi, A. Environmental Impacts and Embodied Energy of Construction Methods and Materials in Low-Income Tropical Housing. *Sustainability* **2015**, *7*, 7866-7883.

1. Introduction

Extensive research has been conducted to improve the energy efficiency of domestic and non-domestic buildings in countries with cold or moderate climates where space heating dominates the energy requirements. The attention is gradually shifting towards tropical climates where cooling governs the energy requirements; however, very little has been done on the low-income tropical housing where operational energy for space heating and/or cooling (excluding cooking and hot water) is negligible. Embodied energy, in contrast, is the key factor for improving the energy efficiency of low-income houses in tropical climates. According to Cabeza *et al.* [1], almost all available studies in the area of Life Cycle Assessment (LCA) and embodied energy have been carried out in developed countries, and there are no case studies in African countries. Ramesh *et al.* [2] also mention that most available studies on LCA are from "cold countries" where space heating is the major concern. There is therefore a clear gap in the literature regarding the environmental impacts and embodied energy of construction methods and materials in African countries.

Located in East Africa, Uganda is one of the developing countries with a moderate tropical climate [3,4] and a high rate of low-income housing. Around 38% of Uganda's population live below the international poverty line [5,6] and over 60% [7] of the Ugandan urban population live in slums with poor quality and hygiene conditions. According to the UN-Habitat, 6 million Ugandan households live in 4.5 million residential units, which means an average occupancy density of around 1.3 families per dwelling. The conditions are more critical for slum residents, the majority of whom live in single-room rented properties [8]. The main fuel sources for cooking and lighting in Uganda are wood fuel and paraffin, respectively. Access to electricity is very limited particularly in rural areas. In 2011, for instance, less than 15% (5.3% in rural areas and 55.4% in urban areas) of the households had access to electricity [9].

The housing conditions in Uganda have been deteriorating due to the lack of proper building regulations and the absence of authorities to enforce minimum housing standards [10]. The failure of the country to adopt low-cost building technologies [10] to deal with low-income housing has also been contributing to the current situation. Introducing new and innovative construction methods and materials which are adapted to local needs and conditions may improve the current conditions. It should be noted, however, that innovation is an evolutionary process, and it is not correct to assume that replacing the current construction methods and materials with new highly efficient, and possibly, more expensive ones would improve the current conditions. The experience has shown that ignoring the current conditions, potentials and limitations of developing countries may result in the failure of an innovation [11].

Increasing use of low quality, energy intensive, locally produced materials (e.g., burned bricks) and gradual shift from environmentally friendly materials (e.g., adobe; thatched roof) towards environmentally harmful ones (e.g., concrete; iron sheet roof) are the other major issues which have led to an alarming situation that requires urgent attention. The current situation has negatively affected the local environment contributing to issues such as deforestation, desertification, air pollution, excessive agricultural soil extractions, food and fuel crisis, and health issues in the country [12,13]. It is obvious that the current situation could lead to social and environmental disasters if no positive action is taken to improve the conditions.

Life cycle assessment is a method of evaluating environmental impacts of buildings during their lifetime from the extraction of raw material though to demolition and disposal [14,15]. According to BS EN 15978:2011, the main stages in building life cycle assessment are as follows [16]:

(1) Production stage;
(2) Construction;
(3) Use stage; and
(4) End of life stage.

The production stage consists of three stages of (a) material supply; (b) transport; and (c) manufacturing [16]. Raw material extraction and its impact on the environment as well as manufacturing and waste are some of the key assessment criteria in LCA. The CO_2 emissions from fossil fuels during extraction and the manufacturing process of construction materials form the majority of the embodied energy/carbon of building products [14]. The latter is, in fact, the major issue associated with Ugandan low-income housing that suffers from excessive energy and material waste during production and handling processes of construction materials.

Although operating energy accounts for a much higher portion of CO_2 and greenhouse gas emissions during the building lifecycle [2,15,17] (75%–90% operational compared to 10%–25% embodied carbon [14]), due to the negligible operational energy in Ugandan low-income housing, the embodied energy of building materials is key to evaluating the environmental impacts of the low-income housing sector. To this end, this research intends to evaluate the current conditions of Ugandan low-income housing in order to highlight the environmental damages and embodied energy of construction methods and materials as well as the key areas which should be addressed to improve the conditions and mitigate the environmental impacts of prevailing methods of construction in Uganda.

2. Methodology

Literature review, and primary data gathered from site visits and photographic surveys are the main methods of data collection for this paper. Relevant documents published by individual researchers, Ugandan Government, UN-Habitat and other research organisations were reviewed. Site visits and photographic surveys were also carried out in two East African countries (Uganda and Tanzania) to collect relevant information on prevailing construction methods and materials and on their environmental impacts in rural areas. Both countries have similar construction methods and the included pictures in this document are from either Uganda or Tanzania. The outcomes of the literature review along with the surveys are used to evaluate the current conditions of low-income housing in rural areas. The embodied energy values of walling methods and materials are also calculated and compared with standard construction methods using the available data in the "Embodied Carbon: The Inventory of Carbon and Energy" [18]. The key factors for improving energy efficiency and reducing the environmental impacts of the low-income housing sector are highlighted and recommendations are provided as the conclusion of this paper.

3. Construction and Housing Conditions

The construction industry is one of the key sectors of Ugandan economy. Different sources suggest different GDP shares for the construction industry in Uganda. According to Oling *et al.* [19], the share of the construction industry in Uganda increased from 13.4% in 2008 to 14.6% of GDP in 2012. This is while, according to Uganda Bureau of Statistics [20], the share of the construction industry in 2011 was 13.1% (Figure 1) showing an increase of 0.8% over 2008 figures. Nevertheless, housing shortages are a major issue facing the Ugandan construction industry. The housing shortages in 2005/6 and 2010 were estimated to be 560,000 [21] and 612,000 [10] units, respectively. The housing deficit is estimated to rise to around 885,000 by 2020 and eight million units by 2040, one million of which would be in Kampala [21].

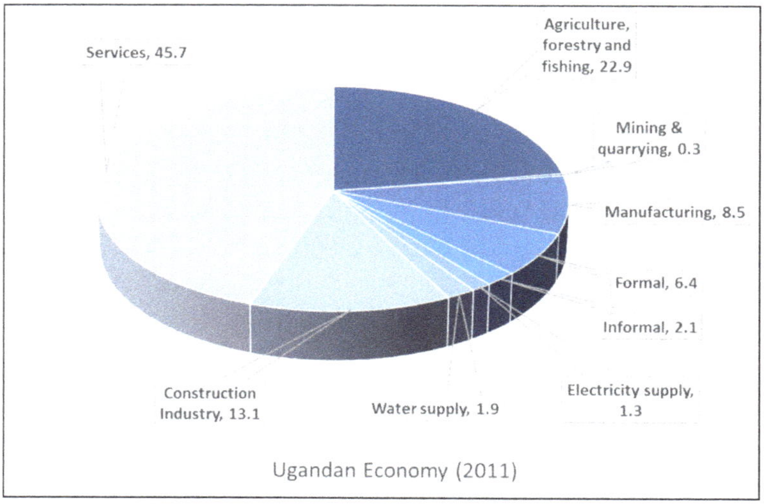

Figure 1. Ugandan Economy in 2011 (%). Source of table: [20].

Housing developers in Uganda follow the market forces rather than the country's actual requirements in terms of affordability [22]. The increasing gap between the supply and what people can afford has pushed the prices up and has forced many people out of the housing market resulting in more demand for rental housing. In 2009/10, for instance, 18% of the population were renting their properties. The situation is more critical in urban areas as, for example, around 70% of households in Kampala live in rental properties [23]. Moreover, incremental self-building by both poor and wealthy people [8], in addition to the lack of quality controls by the authorities [10], have resulted in poor housing quality and have made the housing industry less attractive to potential investors [21]. The situation is more critical in rural areas where the share of good quality housing is much less than in the urban areas [21]. Figure 2 shows the most common types of housing and tenancy types in Uganda in 2009/2010.

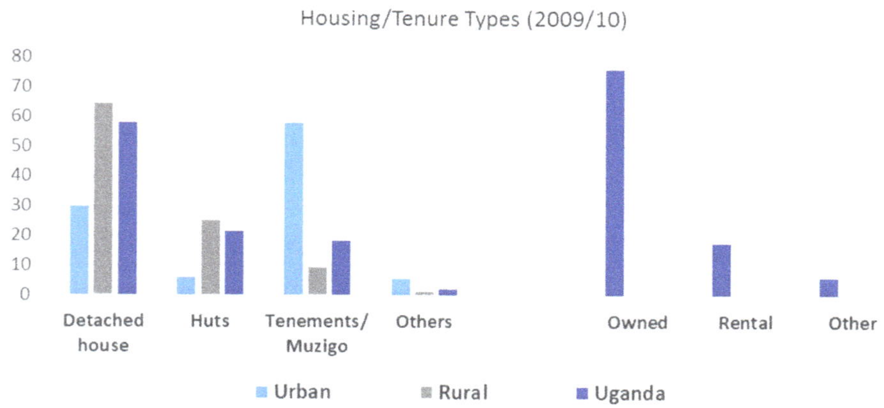

Figure 2. Types of dwelling and tenure in Uganda (2009/2010). Source of table: [23].

Overcrowding is another major issue in Uganda. The average number of occupants in many houses, particularly in urban areas, is more than the international standards of two persons per room [4]. The national average household size in 2011 was 4.9 [9] and the average number of occupants per room in 2009/2010 was 2.9 persons [23]. According to the Uganda National Household Survey 2009/2010, 33% of Ugandan families use two rooms for sleeping and 44% use only one room [23]. The situation is more critical in Kampala where around 70% of houses use only one room for sleeping. Figures 3 and 4 summarise the average number of bedrooms, household size and sleeping occupants in each room in different regions of Uganda in 2009/2010.

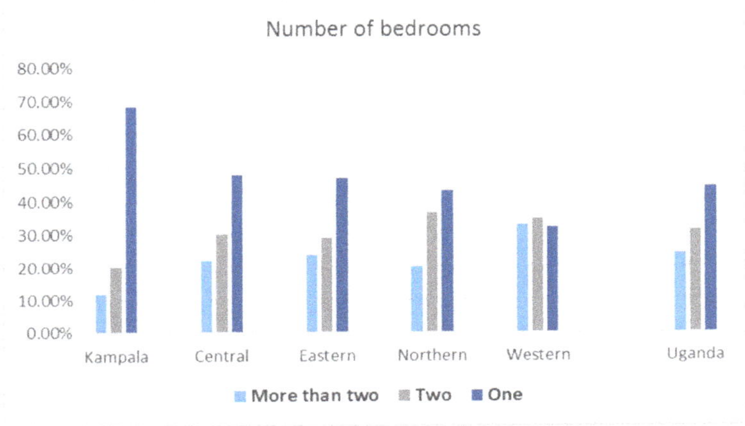

Figure 3. Average number of bedrooms (2009/2010–2011). Source of table: [9,23].

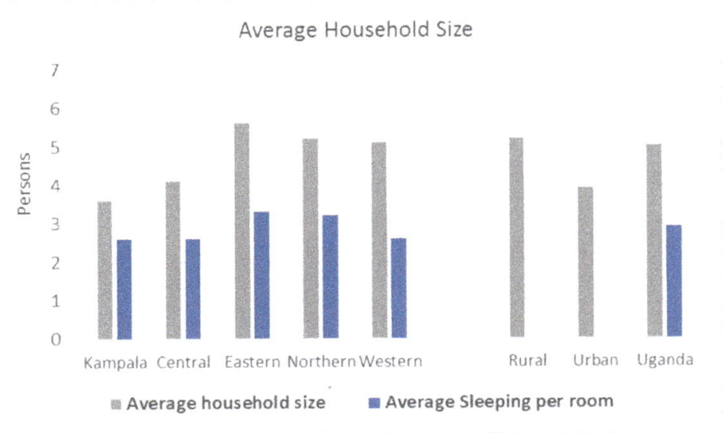

Figure 4. Average household size and sleeping occupants in each room (2009/2010–2011). Source of table: [9,23].

4. Construction Methods and Materials

The major limitations of the construction industries in African countries are high costs and limited access to good quality materials whether imported or produced locally. Moreover, due to the weak transportation infrastructure and limited access to urban factories, particularly in rural areas, material supply cannot cope with the high housing demand [24]. Using locally available materials seems to be an appropriate solution which has already been considered in many parts of Uganda. Figure 5 shows the main construction materials in 2009/2010 in Uganda. Brick and mud and poles are the two most common walling materials/methods in Uganda. Brick walling is the most common system in both urban and rural areas; however, mud and poles walls are much more common in rural than in urban areas. The walling methods are explained in more detail in the following sections.

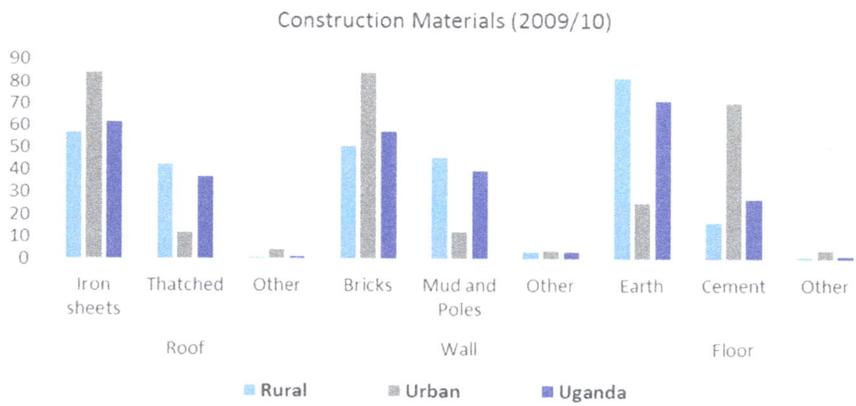

Figure 5. Main types of construction materials in 2009/2010 (%). Source of table: [23].

Iron sheets and thatched roofs (Figure 6) are the most common roofing systems in Uganda. Around 62% of all roofs in Uganda are covered with iron sheets compared to around 37% which are thatched roofs. Iron sheet roofing is much more common in both rural and urban areas. Around 84% of roofs in urban areas are covered with iron sheets compared to 57% in rural areas. This figure for thatched roofs is 43% in rural areas compared to 12% in urban areas. Earth flooring is dominant method/material particularly in rural areas. More than 70% of all floors in Ugandan homes are made out of earth. The share of rural areas is 82% compared to around 25% for urban areas. Cement/concrete flooring is the next most common flooring methods in Uganda. Nearly 30% of all floors in Ugandan homes are made from cement. The share of cement flooring is much higher in urban areas. More than 70% of floors in urban areas are made out of cement/concrete compared to around 17% in rural areas [23].

Figure 6. Iron sheet roof (**left**) and thatched roof (**right**). Source: The authors.

5. Environmental Impacts of Walling Methods and Materials

Many of the prevailing construction methods and materials in Uganda are environmentally harmful. This is because of the inefficient, wasteful and energy intensive production and construction processes. Considering the negligible operational energy of the low-income housing sector in Uganda, embodied energy is the key factor in evaluating the sustainability of low-income housing; however, other issues such as deforestation, air pollution soil extraction, *etc.* are also important issues in evaluating environmental impacts of construction methods and materials in low-income East African housing. This section intends to assess the adverse environmental impacts of three common walling methods/materials in Uganda as follows: (1) Mud and Poles; (2) Bricks (burned and sun-dried/adobe); and (3) Concrete.

5.1. Mud and Poles

Mud and poles is a very common construction method in Uganda (Figure 7). In 2010, more than 39% of all dwellings in Uganda were made with mud and poles. This figure was 46% for houses in rural areas compared to 12% in urban areas [23]. Mud and poles construction consists of wooden structure (poles) and the space/gap between the poles is filled with mud. Thanks to the local availability of its materials, mud and poles is very economical, making it suitable for low-cost housing; however, it is structurally weak and needs constant maintenance as it cracks easily due to the shrinking/swelling of the wood [25]. Although mud and poles construction can be considered as a sustainable environmentally friendly method of construction (thanks to the application of local unprocessed materials and its negligible embodied energy), it contributes to deforestation due to the use of local wood in its structure [12]. This makes mud and poles construction less environmentally friendly compared to other methods and materials such as adobe, explained below.

Figure 7. Cob (**left**); Mud and Poles (**right**). Source: The authors.

5.2. Bricks

Brick walling is the most common construction method in Uganda. Bricks (either burned or adobe/sun-dried bricks) are readily available in both urban and rural areas of Uganda. Nearly 60% of all houses in Uganda have brick walls [23]; however, it is not clear as to what percentage of walls are made out of burned bricks and/or adobe.

5.2.1. Burned Brick

Burned brick is considered as a durable material; however, it is environmentally harmful due to its low quality, very inefficient production processes and the use of local wood in brick kilns which contribute to deforestation and air pollution [12,13] (Figure 8).

Figure 8. Inefficient production processes of burned bricks and use of local wood contribute to deforestation. Source: The authors.

Excessive use of mortar (Figure 9) during construction is another major issue associated with burned brick walls. According to Perez [12], this is due to the uneven sizes of burned bricks; however, considering disproportionate mortar thicknesses of up to 30 mm, it seems that other reasons such as poor construction skills also contribute to this phenomenon. The unpleasant appearance of brick walls also necessitates plastering [12]. These issues may increase the overall cost of brick walling, making it rather inappropriate for low-cost housing. It is therefore necessary to improve the quality, production and construction processes of burned bricks not only to reduce the overall costs but also to mitigate their environmental impacts.

Figure 9. Excessive use of mortar increases the environmental damage as well as the overall construction costs. Source: The authors.

Alternative construction methods and materials such as interlocking blocks/bricks (Figure 10) have been developed to reduce/eliminate the use of mortar in the construction processes of brick walls. Interlocking bricks, however, are not structurally stable against horizontal and/or earthquake forces. Vertical reinforcement elements may be used to improve their resistance against earthquakes [25].

Figure 10. Interlocking bricks can be an alternative for burned bricks. Source: The authors.

Some interlocking soli blocks are cheaper (per square metre) than burned bricks [12,13] making them very suitable for low-income housing. However, production of stabilised soli blocks is slower than normal bricks; although their onsite assembly is quicker due to the mortar-less construction. Moreover, they require special pressing equipment which may not be readily available in some rural areas. Stabilised soli blocks also require constant monitoring and supervision [26] to maintain their quality during production.

5.2.2. Adobe

Thanks to its very low embodied energy, adobe is an environmentally friendly construction material (Figure 11). Adobe, however, is not as structurally stable and durable compared to burned bricks and requires regular maintenance. The greatest disadvantage of adobe is its vulnerability to water and rain [24]. Internal/external plastering/rendering may improve the appearance and performance of adobe walls reducing the need for maintenance [25]. Despite its environmental advantages, adobe is locally regarded as the "material of the poor" [13], which may restrict its application in Uganda. Educating people about the environmental benefits of adobe and other similar construction materials may help to increase the share of these materials in East African countries. Improving the quality and structural stability of such methods and materials such as mud and poles and adobe to reduce the extent and frequency of maintenance would also improve the public attitudes and encourage the public to consider these methods and materials in their building.

Figure 11. Adobe is one of the most environmentally friendly construction materials. Source: The authors.

The excessive extraction of fertile agricultural soil is a negative issue associated with bricks and similar construction materials (Figure 12). Extensive and excessive soil removal may also affect people's health by creating pools and still water which provides suitable environment for mosquitoes and malaria [12]. This also applies to other local construction methods and materials such as mud and poles, which heavily rely on the earth as their major source of raw material. However, compared to burned bricks, adobe and mud and poles are less environmentally damaging as they do not need firewood in their production processes.

Figure 12. Excessive soil extraction damages agriculture and may affect people's health. Source: The authors.

5.3. Concrete

The local production and consumption of cement in Uganda have been increasing during the last decade. The local production of cement and its net domestic supply in 2011 respectively increased by around 12% and 24% compared to 2010 [20]. Figure 13 shows the cement production, consumption, imports and exports during 2007–2011. Living in concrete homes represents a higher social status in Uganda (Figure 14); however, there are concerns over the thermal comfort of the occupants living in concrete homes due to its weak thermal resistance (Table 1) [12]. This requires more investigation to evaluate the effects of different construction methods and materials on the thermal comfort conditions. Moreover, relatively higher costs of concrete [12] may raise questions over its suitability for low-income housing.

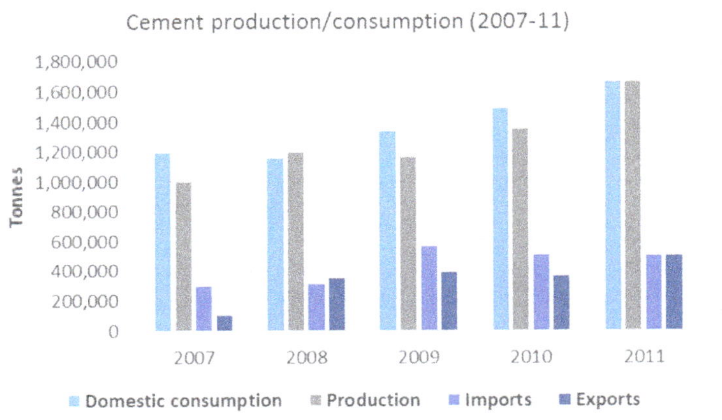

Figure 13. Cement production/consumption during 2007–2011. Source of table: [20].

Although concrete is strong and durable, it is considered as an environmentally harmful material due to high-energy consumption during the production of cement. Nevertheless, compared with burned bricks, concrete may be regarded as a more sustainable material considering the inefficient and energy wasteful production processes of burned brick, which makes it rather more environmentally damaging. Moreover, durability and longer lifespan of concrete compensate for its embodied energy when considering the overall lifecycle of buildings. Therefore, it could be argued that increased use of concrete is not necessarily a negative issue in Uganda. The next sections explain the embodied energy of brick and concrete walling in more detail.

Table 1. Physical properties of typical construction materials in Uganda. Source: [12].

Material	Interlocking Stabilised Soil Block	Sun-dried Mud Block	Burned Clay Brick	Stabilised Soil Block	Concrete Blocks
Dimension: L × W × H (mm)	265 × 140 × 100	250 × 150 × 70 to 400 × 200 × 150	200 × 100 × 100	290 × 140 × 115	400 × 200 × 200
Thermal conductivity (W/m.K)	0.8–1.4	0.4–0.8	0.7–1.3	0.8–1.7	1–1.7
Density (Kg/m^3)	1700–200	1200–1700	1400–2400	1700–2200	1700–2200

Figure 14. The share of concrete construction is increasing rapidly. Source: The authors.

6. Embodied Energy of Burned/Fired Bricks

Artisan producers, small-scale and medium-scale manufacturers are the three major brick suppliers in Uganda [27]. According to the World Bank [27], an average of 0.5 m^3 of Eucalyptus wood is required to produce one tonne of artisan clamp fired bricks. Assuming a lower heating value of 17 MJ/kg [28] and a density of 0.56 g/cm^3 [29] for Eucalyptus wood, an average of 4760 MJ ($0.5 \times 560 \times 17 = 4760$) is required to produce one tonne of artisan bricks.

For the small-scale brick manufacturers, the firing period is considerably longer resulting in much higher consumption of wood. The average wood consumption for small-scale manufactures is 1.8 m^3 [27] which means a massive embodied energy of 17,136 MJ per tonne of bricks ($1.8 \times 560 \times 17 = 17,136$). The wood consumption for medium-scale brick manufacturers considerably reduces to around 0.2–0.3 m^3 per tonne of bricks [27] meaning an embodied energy of around 1900–2850 MJ/tonne, which is six to nine times less than the small-scale brick manufacturers.

According to the Inventory of Carbon and Energy, the embodied energy of "General Clay Bricks" is 3000 MJ/tonne [18]. Therefore, the energy consumptions by artisans and small-scale brick producers are respectively 1.6 and 5.7 times more than the average energy required for brick production in developed countries. It should be noted that the 5.7 times energy consumption may be a better representative of the actual conditions in East African countries. This is because artisans keep the firing period and temperature low to save as much fuel as possible, which in turn results in low quality bricks up to 45% of the entire production. Improving moulding, kiln construction and production processes could save up to 75% in energy consumptions during brick manufacturing [27]. Table 2 summarises the fuel consumption, embodied energy and potential energy saving rates of bricks produced by artisans, small- and medium-scale manufacturers.

Table 2. Embodied energy of burned/fired bricks.

Production Scale	Required Fuel (m^3 of Equivalent Fuelwood/Tonne) *	Energy Consumption (MJ/tonne)	Energy Consumption Compared to General Clay Bricks	Potential Savings in Energy through Different Measures *
Artisans	0.5	4760	159%	35%
Small-scale	1.8	17,136	570%	75%
Medium-scale	0.3	2856	−5%	30%

* According to information in [27].

7. Embodied Energy of Walls

Various sizes have been suggested for burned bricks in East African countries. According to World Bank [27], the length, width and height of artisan clamp fired bricks and blocks may vary greatly (220–295 mm (length); 100–150 mm (width); and 65–130 mm height). The weight of bricks may also vary between 2.5 and 7.6 Kg. The embodied energy of clamp fired brick walls for lower and higher size ranges are calculated in Table 3. An average of 20 mm, 1:4 cement mortar with an embodied energy of 1.1 MJ/Kg [18] is assumed for both artisan and small-scale brick and block walls. As described above, excessive mortar thicknesses of up to 30 mm is normal in Uganda.

According to the results, the embodied energy of 150 mm and 220 mm artisan brick walls are 810 and 1067 MJ/m^2, respectively. This is while the embodied energy values of 200 mm hollow concrete block and 140 mm cement stabilised soil block walls (with an embodied energy of 0.59 MJ/Kg and 0.68 MJ/Kg [18]) are 127 MJ/m^2 and 176 MJ/m^2, respectively. Therefore, the embodied energy of artisan clamp fired brick walls is in order around 8.4 and 4.6 times more than concrete and cement stabilised soil block walls with similar wall thickness.

Obviously these figures are much higher for small-scale manufactured fired bricks with an embodied energy of 17.136 MJ/Kg, which is 3.6 times higher than the bricks produced by artisans. Assuming the same brick density, dimensions and mortar thickness, the embodied energy of one square metre of small-scale burned brick wall would be 3542 MJ/m^2. This is 3.3 times more than artisan brick walls and around 28 times more than hollow concrete block walls. Moreover, the embodied energy of a 215 mm solid wall built with General Clay Bricks is 791 MJ/m^2, which is around 26% and 78% less than a 220 mm artisan and small-scale brick wall, respectively.

Table 3. Embodied energy (EE) of brick and concrete walls.

Product	Brick/Block Size: L × W × H (mm)	External Wall Thickness (mm)	EE of Material (MJ/Kg)	Mass per Item/Littre (Kg)	EE of Wall (MJ/m^2)	EE of Wall (MJ/m^3)
Artisan Burned Brick/Block	300 × 150 × 130	150	4.76	7.6	810	5398
	20 mm mortar		1.11	1.65		
	220 × 110 × 65	220	4.76	2	1067	4849
	20 mm mortar		1.11	1.65		
Small-Scale Brick	220 × 110 × 65	220	17.136	2	3542	16,100
	20 mm mortar		1.11	1.65		
General Clay Brick	215 × 102.5 × 65	215	3	2	791	3677
	10 mm mortar		1.11	1.65		
Cement Stabilised Soil Blocks	290 × 140 × 115	140	0.68	8	176	1257
	10 mm mortar		1.11	1.65		
Hollow Concrete Block *	400 × 200 × 200	200	0.59	14	127	636
	20 mm mortar		1.11	1.65		

* Assumptions: 50% Hollow; 8 Mpa compressive strength.

It should be noted that despite much lower embodied energy of concrete blocks compared to burned bricks, the cost of hollow concrete block walls per square metre is around 36% more than artisan clamp fired brick walls [12]. Therefore, it is necessary to reduce the overall costs of concrete construction in order to encourage using concrete blocks in low-income housing. Cement stabilised soil blocks, in contrast, are around 18% cheaper [12] while their embodied energy is considerably lower than burned bricks. Cement stabilised soil blocks are therefore an appropriate environmentally friendly option for low-income housing.

8. Conclusions

This paper briefly discussed the current housing conditions and prevailing methods of construction in Uganda. Unacceptable housing conditions and low quality, energy-intensive construction methods and materials were highlighted as the major issues of the low-income housing sector. Excessive energy waste during the production processes of building materials such as burned bricks, which also contributes to deforestation, air pollution and other environmental issues, is the major concern which should be addressed. Reducing the embodied energy of locally produced materials is therefore key to mitigating the environmental damages of East African low-income housing.

In fact, some of the locally produced construction methods are much more environmentally damaging than, for example, concrete, which is broadly regarded as an energy intensive and environmentally harmful material in the developed world. According to the results, the embodied energy values of burned bricks produced by artisan and small-scale manufactures are 4760 MJ/Kg and 17,136 MJ/Kg which are 1.6 and 5.7 times higher than generic fired clay bricks, respectively. Moreover, the per square metre embodied energy rates of brick walls built with artisans' and small-scale manufactures' fired bricks are in order 1.35 and 4.5 times more than generic brick walls with similar thicknesses. Furthermore, the per square metre embodied energy of hollow concrete block walls is around 88% and 96% less than brick walls built with artisans' and small-scale manufacturers' bricks, respectively. This, in addition to the higher quality and durability, make concrete a much more environmentally friendly material/method of construction compared with brick walling in Uganda.

Hence, it could be argued that the current increasing trend in the production/consumption of concrete is not necessarily a negative phenomenon. However, considerably higher costs of concrete blocks compared to artisan burned bricks is a major barrier for broader application of concrete in low-income housing. Cement stabilised soil blocks, in contrast, have a much lower embodied energy and a comparable cost with burned bricks making them an appropriate option for low-income housing.

Nevertheless, it should be noted that the local construction industry provides employment to many people in East Africa. Therefore, despite its environmental benefits, replacing the current materials with more energy efficient, factory produced materials may have some negative social outcomes. Refining the existing production and construction methods to improve the quality and reduce the wastes and embodied energy of buildings and materials would therefore be a better strategy to address both environmental and social concerns. In this respect, providing alternative and more sustainable fuels for brick kilns (to reduce deforestation and air pollution), improving moulding systems and energy efficiency during production and construction processes as well as providing dedicated areas for local brick production (to avoid excessive widespread soil extractions) and increasing access to renewable energy sources, particularly in rural areas, are some of the solutions which may be considered to improve the current conditions.

Raising public awareness on environmental issues and educating the stakeholders including local manufacturers, builders and artisans to reduce material waste, replanting trees, and adopting more energy efficient production methods would also help to reduce the environmental impacts of prevailing construction methods and materials in Uganda. Moreover, considering national programmes to change the negative public attitude towards sustainable materials such as adobe would help to increase the share of such materials in the housing industry. Simultaneously, improving construction quality and aesthetics of sustainable methods and materials would make them more appealing to the public and encourage the society to reconsider them as stylish materials which could replace environmentally harmful materials such as burned bricks. National building codes and minimum standards should also be implemented to gradually improve the quality of housing stock in Uganda.

Furthermore, alternative innovative methods and materials which are adapted to local needs and conditions should be developed and introduced to the construction industry. However, it should be noted that innovation is an evolutionary and long-term process and a hasty replacement of the current construction methods and materials with new highly efficient and possibly more expensive ones would probably deteriorate the current conditions. The most effective approach is therefore to improve the production and construction processes of prevailing methods and materials while new and more efficient ones are gradually introduced to the construction industry.

Acknowledgments

This work is funded through an EPSRC research programme, Energy and Low Income Tropical Housing, Grant number: EP/L002604/1.

Author Contributions

Hashemi carried out the literature review, analysed the data and transcribed the paper. Cruickshank and Cheshmehzangi contributed to the design of the research and carried out the site visits and photographic surveys. All authors approved the final manuscript.

Conflicts of Interest

The authors declare no conflict of interest.

References

1. Cabeza, L.F.; Rincón, L.; Vilariño, V.; Pérez, G.; Castell, A. Life cycle assessment (LCA) and life cycle energy analysis (LCEA) of buildings and the building sector: A review. *Renew. Sustain. Energy Rev.* **2014**, *29*, 394–416.
2. Ramesh, T.; Prakash, R.; Shukla, K.K. Life cycle energy analysis of buildings: An overview. *Energy Build.* **2010**, *42*, 1592–1600.
3. Byakola, T. Improving Energy Resilience in Uganda; Helio International, Sustainable Energy Watch, 2007. Available online: http://www.helio-international.org/Uganda.En.pdf (accessed on 7 April 2015).
4. Uganda Bureau of Statistics (UBOS). *2002 Uganda Population and Housing Census*; Analytical Report; Uganda Bureau of Statistics: Kampala, Uganda, 2006.
5. Malik, K. *Human Development Report 2014, Sustaining Human Progress: Reducing Vulnerabilities and Building Resilience*; United Nations Development Programme: New York, NY, USA, 2014.
6. Economic Policy Research Centre (EPRC). *Uganda 2013 FinScope 17 Survey Report Findings, Unlocking Barriers to Financial Inclusion*; Economic Policy Research Centre: Kampala, Uganda, 2013.
7. United Nations Human Settlements Programme (UN-HABITAT). *Situation Analysis of Informal Settlements in Kampala*; Cities without Slums, Sub-Regional Programme for Eastern and Southern Africa, Kivulu (Kagugube) and Kinawataka (Mbuya 1) Parishes, HS/873/06E; United Nations Human Settlements Programme: Nairobi, Kenya, 2007.
8. The Centre for Affordable Housing Finance (CAHF). *Housing Finance in Africa—2014 Yearbook—5th edition*; A Review of Some of Africa's Housing Finance Markets; The Centre for Affordable Housing Finance in Africa: Parkview, South Africa, November 2014.
9. Uganda Bureau of Statistics (UBOS). *Uganda Demographic and Health Survey 2011*; Uganda Bureau of Statistics: Kampala, Uganda, 2012.

10. National Planning Authority (NPA). *National Development Plan (2010/11–2014/15)*; National Planning Authority: Kampala, Uganda, 2010.
11. Hashemi, A.; Hadjri, K. Offsite construction, a potential answer to the Iranian housing shortages. In Proceedings of the Construction Technology and Management (CTM 2014) International Scientific Conference, Bratislava, Slovakia, 9–11 September 2014.
12. Perez, A. *Interlocking Stabilised Soil Blocks, Appropriate earth technologies in Uganda*; HS/1184/09E; United Nations Human Settlements Programme: Nairobi, Kenya, 2009.
13. CRAterre. Earth Architecture in Uganda, Pilot project in Bushennyi 2002–2004. Available online: http://craterre.org/diffusion:ouvrages-telechargeables/download/id/0be68120f38bc83bb6771acb6f5eca90/file/publicationUganda.pdf (accessed on 19 November 2014).
14. Anderson, J.; Thornback, J. *A Guide to Understanding the Embodied Impacts of Construction Products*; Construction Products Association: London, UK, 2012.
15. Asdrubali, F.; Baldassarri, C.; Fthenakis, V. Life Cycle Analysis in the construction sector: Guiding the optimization of conventional Italian buildings. *Energy Build.* **2013**, *64*, 73–89.
16. The British Standards Institution (BSI). *BS EN 15978:2011: Sustainability of Construction Works—Assessment of Environmental Performance of Buildings—Calculation Method*; The British Standards Institution: London, UK, 2012.
17. Sartori, I.; Hestnes, A.G. Energy use in the life cycle of conventional and low-energy buildings: A review article. *Energy Build.* **2007**, *39*, 249–257.
18. Hammond, G.; Jones, C. *Embodied Carbon: The Inventory of Carbon and Energy (ICE)*; BSRIA: Bracknell, UK, 2011.
19. Oling, V.; Rwabizambuga, A.; Warren-Rodriguez, A. Uganda, African Economic Outlook. 2014. Available online: http://www.africaneconomicoutlook.org/fileadmin/uploads/aeo/2014/PDF/CN_Long_EN/Ouganda_EN.pdf (accessed on 20 April 2015).
20. Uganda Bureau of Statistics (UBOS). *2012 Statistical Abstract*; Uganda Bureau of Statistics: Kampala, Uganda, 2012.
21. Kalema, W.; Kayiira, D. Access to Housing Finance in Africa: Exploring the Issues (No.4), Uganda; FinMark Trust, 2008. Available online: http://www.ssauganda.org/uploads/5C2a_Access_to_Housing_Finance_in_Uganda.pdf (accessed on 9 January 2015).
22. Kalema, W.; Kayiira, D. Overview of the Housing Industry and Housing Finance Sector in Uganda; Centre for Affordable Housing Finance in Africa, 2012. Available online: http://www.housingfinanceafrica.org/wp-content/uploads/2012/10/Developments-in-Ugandas-Housing-Finance-Markets2.pdf (accessed on 17 November 2014).
23. Uganda Bureau of Statistics (UBOS). *Uganda National Household Survey 2009/10*; Uganda Bureau of Statistics: Kampala, Uganda, 2010.
24. Batchelder, D.; Caiola, R.E.; Davenport, S.W. *Construction Reference Manual, A Source Book for the Use of Local Materials in Construction*; The Experiment in International Living: Brattleboro, VT, USA, 1985.
25. Minke, G. *Construction Manual for Earthquake-Resistant Houses Built of Earth*; GATE–BASIN (Building Advisory Service and Information Network): Eschborn, Germany, 2001.

26. Engineering Ministries International (EMI). *Architectural Design Guide*; Engineering Ministries International—East Africa: Kampala, Uganda, 2012.
27. World Bank. *Uganda—Energy Efficiency Improvement in the Brick and Tile Industry*; Activity completion report; no. ESM 97 89, Energy Sector Management Assistance Programme; World Bank: Washington, DC, USA, 1989.
28. Musinguzi, W.B.; Okure, M.A.E.; Wang, L.; Sebbit, A.; Løvås, T. Thermal characterization of Uganda's Acacia hockii, Combretum molle, Eucalyptus grandis and Terminalia glaucescens for gasification. *Biomass Bioenergy* **2012**, *46*, 402–408.
29. Kumar, R.; Pandey, K.K.; Chandrashekar, N.; Mohan, S. Study of age and height wise variability on calorific value and other fuel properties of Eucalyptus hybrid, Acacia auriculaeformis and Casuarina equisetifolia. *Biomass Bioenergy* **2011**, *35*, 1339–1344.

Profiling Space Heating Behavior in Chilean Social Housing: Towards Personalization of Energy Efficiency Measures

Victor Bunster and Masa Noguchi

Abstract: Global increases in the demand for energy are imposing strong pressures over the environment while compromising the capacity of emerging economies to achieve sustainable development. In this context, implementation of effective strategies to reduce consumption in residential buildings has become a priority concern for policy makers as minor changes at the household scale can result in major energy savings. This study aims to contribute to ongoing research on energy consumer profiling by exploring the forecasting capabilities of discrete socio-economic factors that are accessible through social housing allocation systems. Accordingly, survey data gathered by the Chilean Ministry of Social Development was used identify key characteristics that may predict firewood usage for space heating purposes among potential beneficiaries of the Chilean social housing program. The analyzed data evidences strong correlations between general household characteristics and space heating behavior in certain climatic zones, suggesting that personalized delivery of energy efficiency measures can potentially increase the effectiveness of initiatives aimed towards the reduction of current patterns of consumption.

Reprinted from *Sustainability*. Cite as: Bunster, V.; Noguchi, M. Profiling Space Heating Behavior in Chilean Social Housing: Towards Personalization of Energy Efficiency Measures. *Sustainability* **2015**, *7*, 7973-7996.

1. Introduction

Energy efficiency has become a priority concern for both developing and developed nations. Current patterns of energy consumption can have strongly negative social, economic and environmental impacts; hence the need for policies to further promote a transformation of the way in which natural resources are been harvested, distributed and consumed [1]. In this context, a significant amount of energy is being used in the operation of residential buildings. The International Energy Agency explains that almost 23% of the total energy consumed in 2011 was used in the residential sector, meaning that since 1990 the domestic demand has grown by 35% [2]. Significantly, most of this energy is being used for space and water heating [1], suggesting that household activities and perceptions may have a strong impact on overall patterns of consumption.

Policy-makers have explored different approaches to tackle this issue including building retrofit programs [3], rating systems [4], and information campaigns to promote behavioral changes [5,6]. Among them, those strategies focused on feeding personalized information back to the consumers have been shown to be remarkable efficient and cost-effective [7,8]. However, implementation of these principles is often not an option for developing nations, as gathering information on the recurrent patterns of operational energy of different users use requires centralized information systems that may not be available or may not even account for some of the fuels used for space heating.

This study explores an alternative approach to energy consumption profiling aimed to inform personalized energy efficiency measures under such technologically lagging conditions. Accordingly, the Chilean social housing program is used as research context to assess the potential firewood consumption forecasting capabilities of discrete socio-economic factors that can be considered before occupancy through subsidy allocation systems.

2. Background

2.1. Chilean Social Housing

Chile is at the threshold of becoming a developed nation. Over the past decades, the Chilean economy has been one of the fastest growing in Latin America, becoming the first country of the region to join the Organization for Economic Co-operation and Development (OECD) by 2010 and starting to be officially recognized as a high-income economy by the World Bank in 2013. One of the consequences of this rapid economic growth has been an unprecedented increase in the internal demand for energy. Over the past 20 years the total consumption of primary energy has more than doubled, while the internal demand for electricity almost quadrupled [9]. However, lack of a clear strategic plan for sustainable development of the energy sector has resulted in a high dependence on imported fossil fuels that maintains the country in a position of high vulnerability [10]. In this context, the Chilean Government has progressively introduced programs and institutions to deal with the current consumption rates whilst promoting efficient use of the available resources.

One of the main areas of the Chilean Agency of Energy Efficiency is operational energy consumption of buildings. By 2012 more than 13% of the country's total demand for energy came from the residential sector, 66% of which has been estimated to be used solely for space heating purposes [11,12]. This has motivated a number of energy efficiency initiatives specifically focused on housing, including massive educational campaigns and subsidies for the installation of solar water heaters, compact fluorescent bulbs, and upgrading to slow combustion heaters [11].

In this context, the use of biomass for space heating in social housing has significant economic, environmental and public health implications. During 2012, biomass constituted 28% of the total primary energy consumed in Chile, 12% of which was used for space heating in the residential sector [11,12]. Gomez-Lobo [13] explains that the ubiquitous use of firewood for space heating in southern Chile and the lack of adequate regulation are the main reasons behind the high levels of pollution and increasing degradation of native forests of this region. Significantly, an important amount of this firewood consumption can be assumed to be used in social housing as estimations of the Ministry of Social Development suggest that a significant percentage of the current housing stock was acquired though public subsidy programs [14] and most of the firewood consumed in southern Chile is known to be used by households of the lowest socio-economic segments [15] (Figure 1).

Figure 1. Social housing development in Los Angeles, southern Chile.

Different studies have shown that social dwellings lack minimum standards to ensure adequate thermal comfort levels during the cold seasons [16,17]. In this context, new regulation and subsidies have been introduced aiming to improve the standards of both new (*i.e.*, the Thermal Regulation, Article 4.1.10 in [18], see also [19]) and existing housing stock (*i.e.*, the Thermal Reconditioning Program, Article 6 in [20], see also [21]). Nonetheless, such initiatives may have strong limitations as they are intended to address no more than the minimum conditions for the average household, or to act as palliative solutions for highly inadequate dwellings. Although design has been shown to be a defining factor for the performance of the dwellings [22,23], there is still limited information regarding the impact of different households on their overall performance that could inform future initiatives towards the reduction of operational energy consumption [24].

2.2. Energy Consumer Profiling

Operational energy consumption for space heating is a problem that goes beyond the architectural characteristics of a dwelling. Different end users may have contrasting perceptions of comfort, as well as different priorities and strategies when dealing with uncomfortable situations. In this context, there is a growing body of research focused on energy consumption profiling that may be used to assess general patterns of residential energy usage without the need of on-site measurement or centralized information systems [25–27].

Fredericks *et al.* [25] make a comprehensive review of the available literature to explain that most of the research focuses on the problem of finding end user profiles that are either above or below average (*i.e.*, "energy-wasting" and "energy saving" users). They propose two main categories of end user profiling methods as those focused on socio-demographic factors (e.g., income, age, household size and educational level), and those focused on psychological factors (e.g., beliefs, intentions, attitudes, and social norms), both of which work in combination with situational factors (e.g., existing laws, socio-cultural factors, and the built environment). Fredericks *et al.* [25] argue that the stronger socio-economic predictors are those related with household size and composition, income, occupation and employment type (shifts), dwelling type and size, and geographic location, whilst other variables such as age, gender, dwelling age and ownership status may show lower consistency levels. Similarly, Abrahamse and Steg [26] identify household size, income, and age as the strongest indicators of domestic energy use. Guerra Satin [27], nonetheless, acknowledges the difficulties of establishing strong relationships between predefined household profiles (*i.e.*, "family", "seniors", "high-income couples", and "singles and low-income couples"), their behavioral patterns and domestic energy consumption. Accordingly, this study uses clustering techniques to define household profiles based on socio-economic factors that have been shown to be strong predictors of overall patterns of energy consumption and may be accessible through the Chilean social housing subsidy allocation system.

2.3. Subsidy Allocation Process

The Solidary Fund of Housing Choice (SFHC) [28] is a social housing program that enables people from the first quintile (autonomous household income per capita) in situations of social vulnerability to acquire housing solutions without mortgages. It defines—in combination with other regulatory bodies—the methods to calculate each subsidy, the application process and selection criteria, as well as minimum technical standards for the dwellings [18,29,30]. A key selection criteria is the Social Vulnerability Score of each application, which combines the score of the Social Protection File (SPF) [31] of the household with the Housing Scarcity Score [32] of their current dwelling.

The SPF is the main instrument to allocate any social expenditure under the current legislation. Candia *et al.* [33] explain that the SPF targets vulnerability indicators by assessing relationships between main socio-economic capabilities and needs of the household, as well as the eventual presence of risk factors that may either increase these needs or reduce the effectiveness of their own defense mechanisms, as:

$$SPF = S.E.Capabilities + Income\ /\ S.E.Needs \times Risk\ Factors$$

The socio-economic capabilities factor is calculated by looking at the potential capacity of each household member at working age to generate income, including variables such as labor skills, experience, educational level, and current occupation, while the income factor includes both formal and informal remunerations as well as any other permanent income such as pensions. The socio-economic needs factor corrects the socio-economic capabilities to the number of household members, but also assesses variables such as age, gender, and kinship. The risk factors are calculated using variables such as number of household members that may have physical or mental disabilities, precarious work or location, and health conditions in general [34].

The Housing Scarcity Score adjusts the total SPF with seven variables that assess residential scarcity in the household throughout the application, including eventual presence of secondary households, overcrowding, housing precariousness, water scarcity, lack of sewage or disposal system, housing inhabitability, and participation in the *Chile Solidario* social protection program. A predefined factor is assigned to each one of these variables to be later subtracted from the total household SPF. The chances of obtaining a subsidy are inversely proportional to this weighted total.

Despite being regarded as an effective method for allocating social expenditure, the FPS has been subject to controversy. Candia *et al.* [33] and Herrera *et al.* [34] argue that, as the information is gathered through personal interviews and the responses are not later verified, there is a tendency to distort issues such as household size and disabilities because they may increase the household's chances of receiving social benefits. Nevertheless, the information contained by the SPF may be used to assess general socio-economic conditions of future dwellers.

3. Methods

3.1. Research Data

The study was conducted using the official CASEN 2013 dataset [35], a cross-sectional nationwide household survey developed by the Chilean Ministry of Social Development. The main objective of this survey is to provide periodical information about the socio-economic state of different Chilean households in order to identify problems and evaluate the impact of social policies. The survey is conducted on site and includes seven modules: registry, education, work, income, health, residents, and dwelling. The CASEN 2013 Survey was conducted nationwide between November 2013 and February 2014, covering 64,842 dwellings and 218,491 people of both urban and rural areas. The survey is voluntary, anonymized, and the datasets are open to the public.

3.2. Target Population and Variables

The first stage of the analysis consisted in the identification of households whose characteristics resemble those targeted by the Chilean social housing regulation [28]. Accordingly, only subsidy dwellings that had a total surface area between 41 and 60 m^2 and were acquired without mortgages by people belonging to the first quintile of per capita autonomous income were included, resulting in 2614 cases that were used as the target population (*i.e.*, the SFHC group). The study focused only in urban dwellings. The rest of the cases were used to enable comparative analyses among quintiles as shown in Section 4.1, but were not included in the household consumption profiling analyses discussed in Section 4.3.

In accordance to the SPF formula, 28 of the 600 available variables of the CASEN dataset [35] were included in the analyses, *i.e.*: region, district, zone, gender, age, marital status, head of household, number of household members and nuclei (Registry module); educational level (Education module); occupation or profession, employment category, monetary subsidies, work shifts, and employment type (Occupation module); total autonomous income, total monetary income, and poverty category (Income module); type of permanent or long lasting health conditions, mobility difficulties, and self-assessed health mark (Health module); and overcrowding index, presence of secondary households, dwelling type, wall material, roof material, use of firewood, and annual firewood consumption (Housing module). A detailed description of these variables can be found in [36].

A further four variables were defined over the CASEN dataset. The age segment variable divided individuals in age groups as child (0 to 5), young (6 to 17), young adult (18 to 35), adult (36 to 59), and elder (60+). The occupation category variable divided the occupation or profession variable into qualified occupations (*i.e.*, military, public servants, professional, science and intellectuals, technicians, office employees, sales and service workers, agriculture and fisheries workers, mechanics, and facilities and machinery operators) and non-qualified occupations (other). The climatic zone variable categorized the district variable into 7 thermal zones as defined by MINVU [37]. The disability type variable divided the type of permanent or long lasting health condition variable into two groups as physical and mobility (including deafness, blindness, and muteness), and mental and psychiatric disabilities (including intellectual).

3.3. Statistical Analysis

At a first stage of analysis the whole dataset was sorted by quintile and contrasted to the SFHC group using the variables described in Section 3.2. Then, these variables were divided into 3 categories according to the SPF formula, *i.e.*, household socio-economic capabilities, household socio-economic needs, and household socio-economic risks, whilst others were divided into 2 further categories: dwelling characteristics and firewood consumption. Pearson Bi-variable Correlation was used to identify statistically significant linear relationships across categories, where results r = 0.00 to 0.19 were interpreted as negligible; r = 0.20 to 0.29 as a weakly correlated; r = 0.30 to 0.39 as moderately correlated; r = 0.40 to 0.69 as correlated; and r = 0.7 to 1.00 as strongly correlated. Positive results imply that the correlation occurs in the same direction (*i.e.*, if one variable increases the other one increases, and if one variable decreases the other one also decreases), and negative results imply that the correlation occurs in the opposite direction (*i.e.*, if one variable increases then the other one decreases and *vice versa*).

At a second stage only statistically correlated variables were clustered using the Two-step Cluster Analysis algorithm. As explained by Bacher *et al.* [38], the main objective of cluster analysis is to identify groups of objects that are as similar between them and as different to other groups as possible. They argue that traditional clustering methods can be categorized into two main groups: relocation (e.g., k-means and expectation-maximization clustering) and hierarchical methods (e.g., agglomerative and divisive clustering), none of which is able to identify the optimum number of clusters automatically. The main advantage of two-step clustering method is the incorporation of this capability by pre-processing the data and generating a cluster feature tree, which is later used to explore different clustering options and determine the optima using, in this case, Schwarz's Bayesian Criterion. All the analyses were performed using IBM SPSS Statistics v22.

4. Results and Discussion

4.1. General Demographic Characteristics

The general demographic characteristics of the target households were evaluated under four main categories and contrasted to the total CASEN 2013 dataset. In order to explore similarities and differences between groups, this population was divided into quintiles. These comparative analyses enabled identification of key features of the SFHC group that were later used to guide the definition of household consumption profiles aimed to inform personalised delivery of energy efficiency measures.

4.1.1. Household Size and Structure

The mean household size of the SFHC group was larger than average with 4.77 people per dwelling in contrast to 4.08 of the CASEN total, and more than one person larger than the national average of 3.6 according to 2002 Census data [39] (Figure 2). This is due to a larger concentration of households in the 3–6 people range with more than 80% of the cases, in contrast to more dispersed distributions in Quintiles 1 to 4 (Table 1).

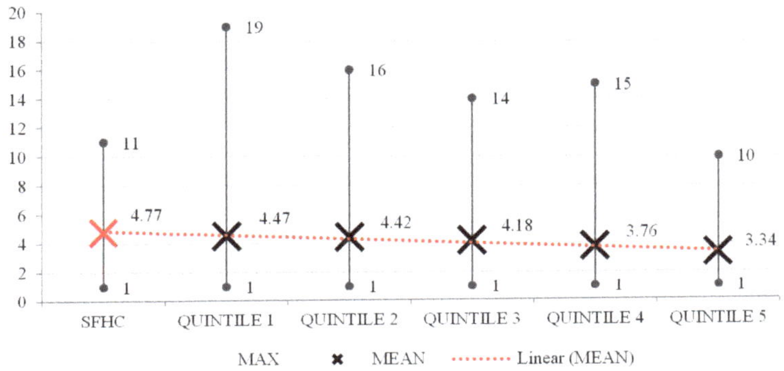

Figure 2. Mean, maximum and minimum number of people per dwelling by quintile contrasted to the Solidary Fund of Housing Choice (SFHC) target population [35].

Table 1. Distribution of occupants per dwelling by quintile and in the SFHC target population [35].

	SFHC	QUINTILE 1	QUINTILE 2	QUINTILE 3	QUINTILE 4	QUINTILE 5
1	1.22%	2.68%	2.15%	3.14%	4.59%	8.44%
2	5.55%	10.36%	10.65%	12.24%	15.71%	22.89%
3	16.88%	18.07%	19.23%	21.85%	26.08%	24.40%
4	26.90%	25.64%	25.86%	25.72%	26.56%	24.74%
5	19.58%	19.20%	19.00%	19.04%	15.25%	12.84%
6	15.87%	11.52%	11.92%	8.77%	7.20%	4.49%
7	5.50%	6.20%	4.29%	4.28%	2.58%	1.20%
8	3.92%	2.82%	4.13%	2.33%	0.92%	0.56%
9	2.93%	1.51%	1.52%	0.88%	0.76%	0.41%
10	1.04%	1.20%	0.87%	1.32%	0.25%	0.03%
11	0.61%	0.31%	0.14%	0.34%	0.03%	0.00%
12+	0.00%	0.48%	0.26%	0.08%	0.06%	0.00%

This tendency towards larger households may have significant consequences for the SFHC group, considering that this variable has been suggested to be among the strongest indicators of domestic energy consumption [25,26]. Significantly, the percentage of overcrowded dwellings (indexes between 2.5 and 3.49) in the SFHC group almost doubled the total average with 18.72% to 8.89% across quintiles. Nonetheless, the ratio of dwellings with medium to critical overcrowding indexes was lower in the SFHC group than in the first quintile (5.78% to 8.66%), but significantly higher than the total average (3.67%) (Figure 3). This can be related to changes to the household structure that happen during the application process, but may also be a result of the overcrowding calculation method (*i.e.*, number of bedrooms/dwellings) as, for example, a single bedroom dwelling with large surface area can have a higher overcrowding index than a smaller one with a larger number of bedrooms.

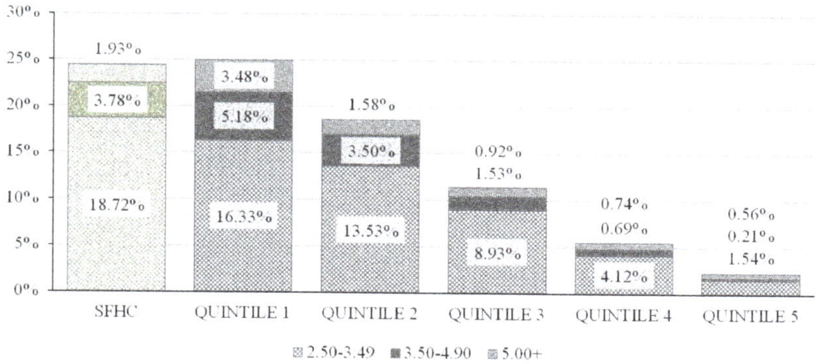

Figure 3. Overcrowding index per quintile and in the SFHC target population [35].

Other possible explanation for this large proportion of overcrowded dwellings might be the number of secondary households found within this group (*i.e.*, two or more households per dwelling). In more than one third of its cases (34.42%) the SFHC group showed secondary households living within the dwelling (*i.e.*, internal, Figure 4), while the number of secondary households living in the same lot but outside the dwelling (*i.e.*, external) was remarkable small (3.53%) when compared to Quintiles 1 to 4 (8.19% to 5.90%). This phenomenon might be linked to the characteristics of the social housing typologies, which are often apartments or units in small lots. These results are also consistent to the structure of the households, as the SFHC group concentrated the larger percentage of binuclear households with 15.96% against 13.21% in the first quintile and 11.63% in the total average, a tendency that also persists in households with three or more nuclei (Figure 4).

In terms of age and gender distribution, the SFHC group showed a strong tendency towards younger population with 19.56% below the adulthood threshold (18 years) in contrast to 12.78% in the total (Figure 5). The proportion of elders (60+ years) was also smaller than average, with 5.75% against 8.35% in the total. A remarkable characteristic of the SFHC group was evident after gender segmentation. While the SFHC female population showed a close to normal distribution (in contrast to the positive skewedness of both male and female population in the total), the SFHC male population showed a significant reduction of individuals at the young adult and adult age segments.

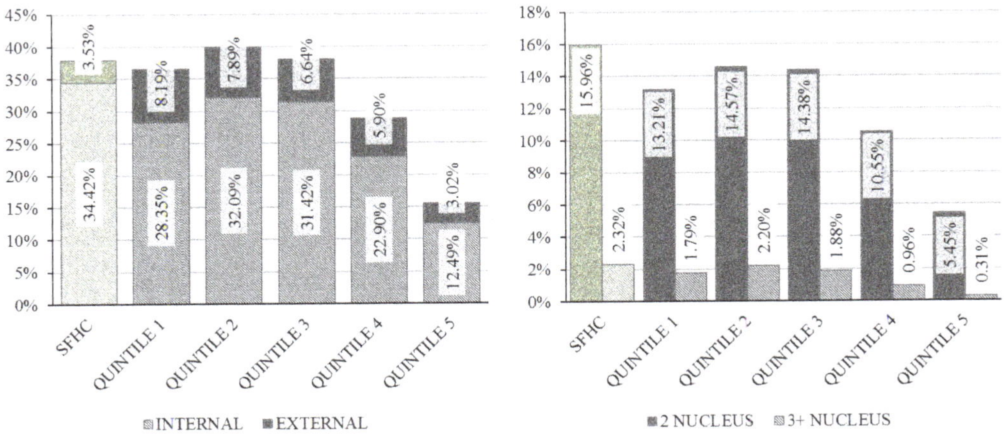

Figure 4. Presence of secondary households (**left**) and number of nuclei (**right**) per dwelling by quintile and in the SFHC target population [35].

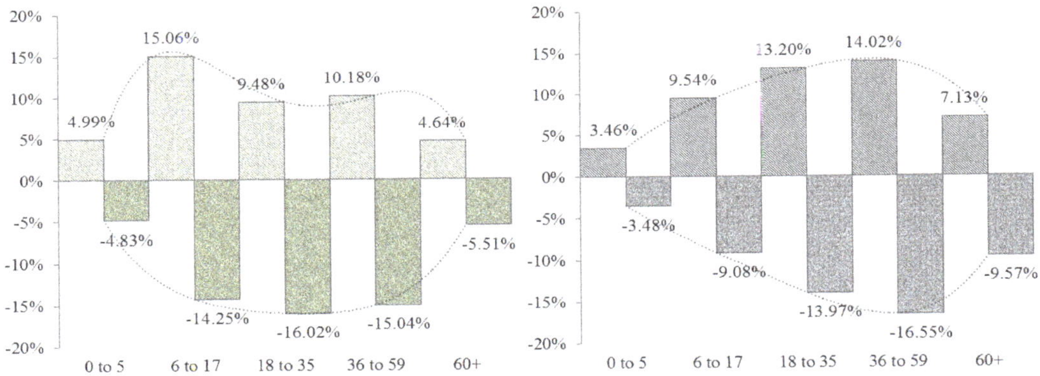

Figure 5. Age segments and gender distribution in SFHC target population (**left**) and total population (**right**); positive numbers show males and negative numbers show females [35].

This tendency towards a larger proportion of female adults and young adults may be related to the amount of female heads of household, which with 53.78% is the higher across quintiles (average of 38.25%) and the only segment where it surpassed the amount of male heads of household with an average of 61.74% (Figure 6). Although the gender of the users has been suggested to be a poor indicator of operational energy consumption [26], the composition of the household may define its evolution over time and therefore its long term patterns of consumption. Accordingly, these patterns may also be related to the marital status of the residents, which in the SFHC group showed a tendency towards larger proportion of singles and the smaller of married and *de facto* couples (Figure 6).

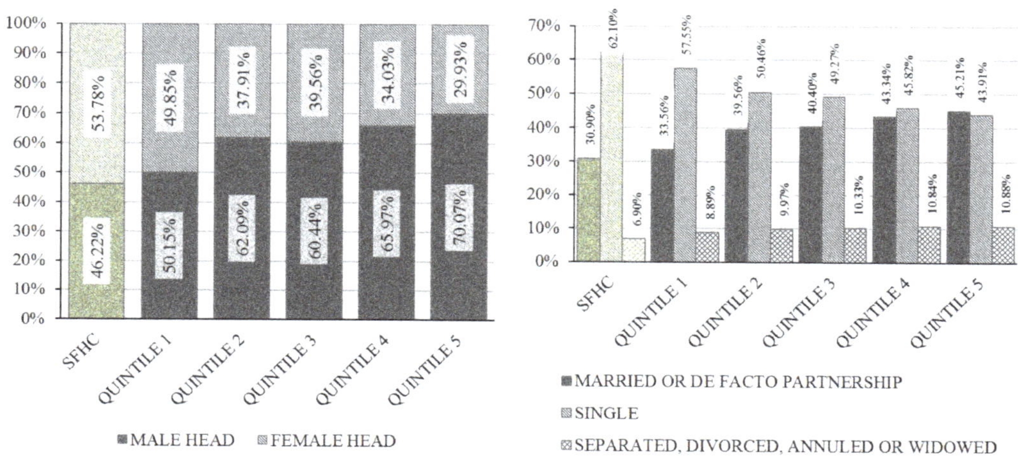

Figure 6. Head of household gender ratio (**left**) and marital status (**right**) by quintile and in the SFHC target population [35].

4.1.2. Education and Occupation

While education has been linked to domestic energy conservation attitudes and perceptions, occupation and work shifts may directly impact the schedules, use of electric artifacts, and investment in energy efficient technologies of a household [25]. The SFHC group showed the largest concentration of individuals with no formal and incomplete primary education (48.97%), and the smaller concentration of individuals that attended tertiary education (0.05%). Both the SFHC group and the first quintile concentrated a large amount of individuals with none or incomplete primary education (48.97% and 46.12% respectively), an issue may be directly linked to the occupation of the individual and therefore to the total income of the household.

In terms of the occupation category, the SFHC group showed the larger proportion of non-qualified workers (63.46%), which are nevertheless negligibly higher than in Quintile 1 (65.22%) (Figure 7). These differences remain negligible in terms of the employment category, as both the SFHC group and Quintile 1 showed the larger proportion of 'minor' workers (*i.e.*, informal) (26.16% and 23.03%) and the smaller percentage of employers or independent workers (6.59% and 7.71% respectively) (Figure 7). Although also negligible, there was a reduction in the number of unoccupied individuals (−4.51%) and an increase in the number of employees (+2.5%).

Figure 7. Distribution of the main occupation by qualification level (**left**) and category (**right**) by quintile and in the SFHC target population [35].

In terms of work shifts, the SFHC group showed a significantly higher concentration of night shifts (5.2% to 1.57%) and a smaller concentration of rotating shifts (9.93% to 13.63%) than the total average (Figure 8). Similarly, in terms of employment type the SFHC group showed the lower rates of full-time employment (76.23% to 84.25%) and the higher ones of part-time workers (17.59% to 9.52%) (Figure 9). These tendencies may have a strong impact on the energy consumption patterns of different households as they are directly linked to their schedules.

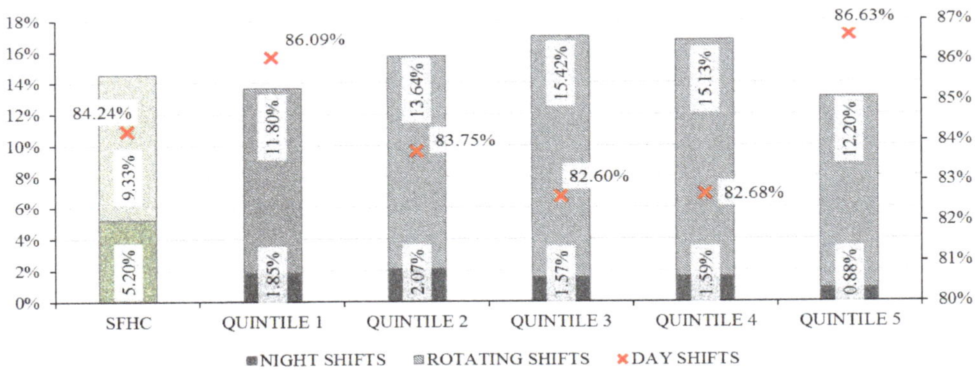

Figure 8. Distribution of work shifts by quintile and in the SFHC target population [35].

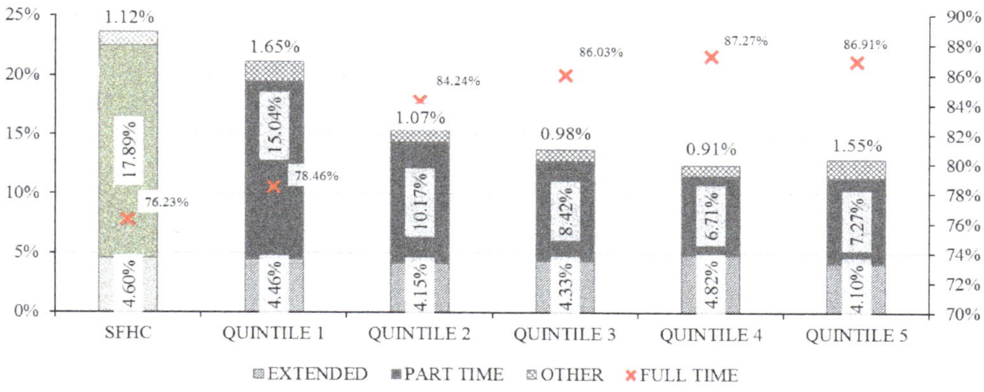

Figure 9. Distribution of employment type by quintile and in the SFHC population [35].

4.1.3. Income and Poverty

Household income is often positively correlated to energy consumption, but may also increase the probabilities of a household to investment in energy efficient technologies [25]. These capabilities may be far from SFHC group, which mean autonomous income was significantly below the national average, both its total household monetary income and per capita monetary income showed minor differences when compared to Quintile 1, and its levels poverty remained significantly higher than average (60.11%) (Figure 10). Significantly, a combination of high consumption rates and low income often result in fuel poverty. However, in the SFHC group these tendencies may be mitigated by monetary subsidies perceived by the households, which almost doubled the ones of the first quintile and were significantly higher than the total average (Figure 10).

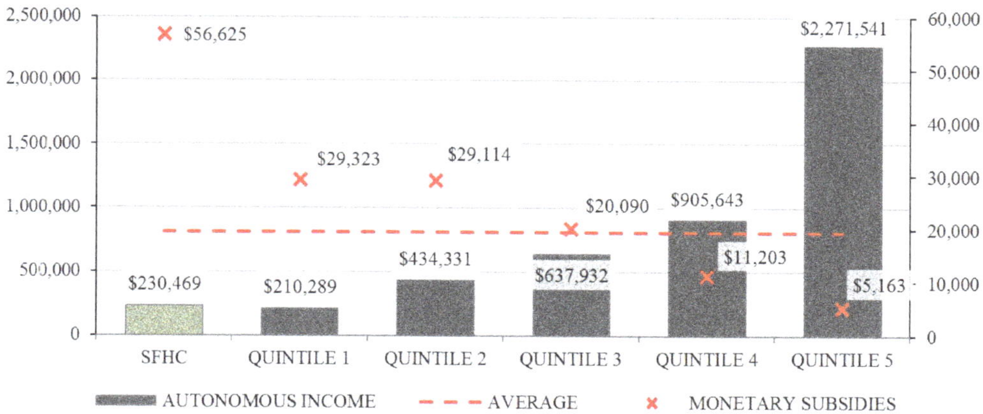

Figure 10. Autonomous income and monetary subsidies in Chilean pesos by quintile and in the SFHC target population [35].

4.1.4. Disabilities and Health Issues

Households with one or more disabled members are more likely to perceive a lowest income, and consequently to fall into fuel poverty [40]. The number of dwellers with disabilities in the SFHC group was smaller than in Quintile 1 (9.75% to 10.74%), but still higher than the total average (6.04%) (Figure 11). This concentration of disabled individuals in the SFHC group may be related to the opportunistic behavior described by Candia *et al.*, [33] and Herrera *et al.*, [34]; however, the analyzed data showed no evidence of these variables having a direct impact on the SPF focalization process. In terms of the self-assessed health score, the data showed higher results in the SFHC group when compared to Quintile 1 (5.64 and 5.56 respectively), reaching the same score than Quintiles 2 and 3 (Figure 11). Although this variable depends upon highly subjective criteria, such an increment might reflect levels of satisfaction among dwellers.

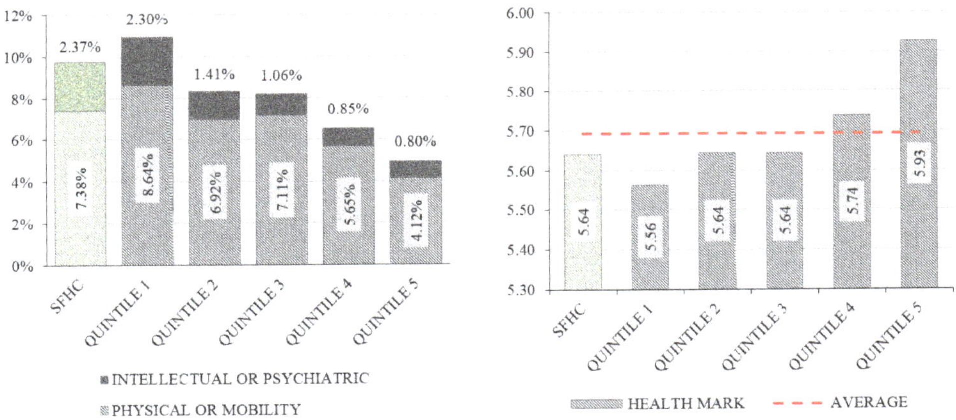

Figure 11. Percentage of people with disabilities by type (**left**) and mean declared health mark by quintile and SFHC target population (**right**) [35].

4.2. Dwelling Characteristics and Firewood Consumption

Among other variables, the space heating behavior highly depends upon material and morphological characteristics that define the basic thermal performance capabilities of a dwelling. The CASEN 2013 survey collects only a limited amount of these variables including surface area, architectural type and envelope materials. These characteristics were analyzed by climatic zone to explore existing relationships among dwellings, environment, users, and energy consumption rates.

4.2.1. Architectural Types

A total of 33.2% of the SFHC cases were living in detached houses at the moment of the survey, while 38.78% were living in semi-detached houses, 10.41% in terraced houses, and 17.32% in apartments with no lifts. No shanties, “mediaguas”, or other precarious settlements were found, suggesting that at the moment of the survey most of the selected cases were living in formal dwellings. Also, whilst most of the population of Quintiles 1, 2, 3 and 5 were living in detached housing types, the SFHC group showed a stronger tendency towards social housing types such as semidetached houses and apartments with no lifts (*i.e.*, "blocks").

Regarding the materials of the SFHC dwellings, the analyzed data suggested a gradual transition from masonry at the warmest climatic zones to lined wood frame in colder areas (Figure 12). Zone 3, the most populated of the country, showed a strong prevalence of masonry with 94.5% of the cases; while 100% of the cases in the colder Zone 6 were built with lined wood frames. Other construction techniques such as adobe and reinforced concrete, which are generally associated to vernacular constructions and mid- to high-rise buildings respectively, have a reduced share of the cases in warmer areas. In terms of the roof materials, there was a strong prevalence of metallic sheets across climatic zones (Figure 12). This may be related to the generalized use of zinc-aluminum sheets in social housing for being one of the most inexpensive, durable, and easy to install materials allowed by normative [28]. No information regarding the thickness or insulation properties of these materials was included in the survey.

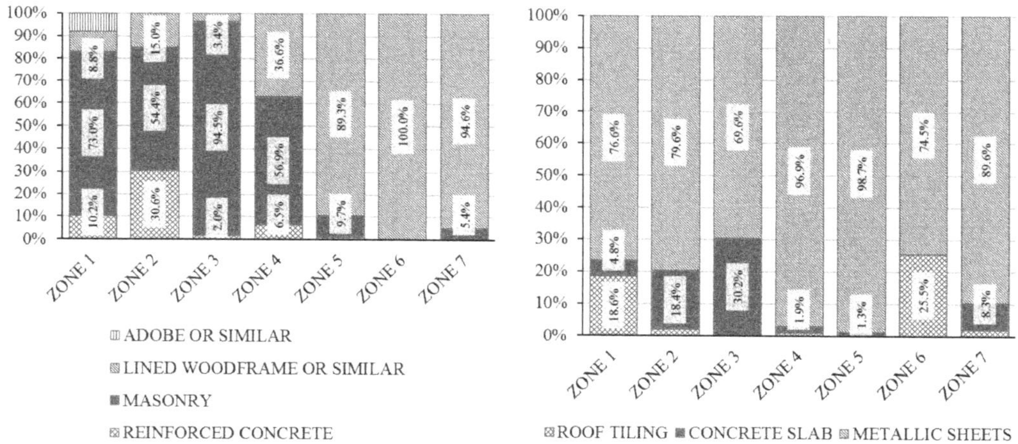

Figure 12. Main building materials of walls (**left**) and roofs (**right**) by climatic zone [35].

4.2.2. Firewood Consumption

The SFHC group confirmed the strong use of firewood for space heating purposes described by the literature [15]. Although the percentage of people that declares its use was larger in the first quintile, when annual kilograms of firewood consumption was compared among quintiles both groups were well above average (Figure 13).

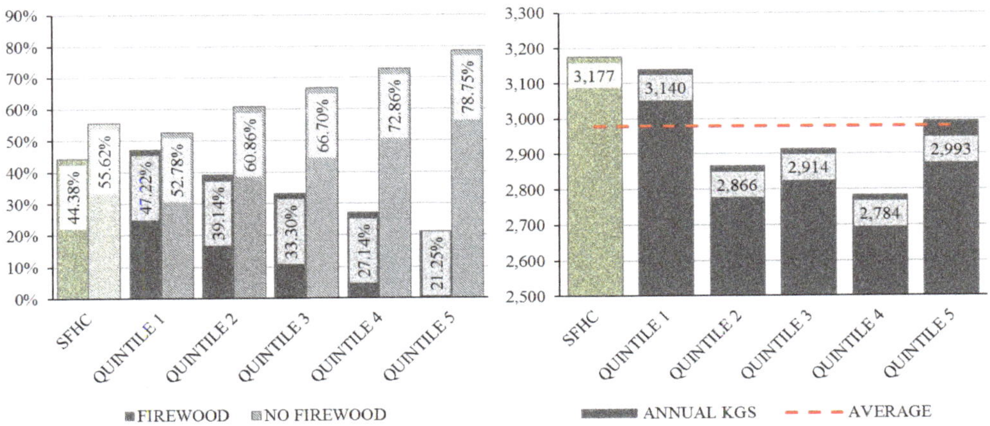

Figure 13. Use of firewood for space heating (**left**) and total annual consumption in kilograms (**right**) by quintile and in the SFHC target population [35].

In parallel, when both variables were analyzed by geographic region the data evidenced a strong increase of firewood consumption in central-southern and southern areas of Chile as described by the literature [14,16] (Figure 14). While the null firewood consumption shown in regions I, II and III may be related to both cultural and climatic conditions, in region XII this may be a result of the ubiquitous use of gas for space heating purposes. Similarly, when analyzed by climatic zone the consumption gradually increased in colder areas, while the reduction in percentage of declared consumption found in Zones 3 and 7 may be an effect of the use of alternative fuels and firewood restrictions in highly populated urban areas (Figure 15).

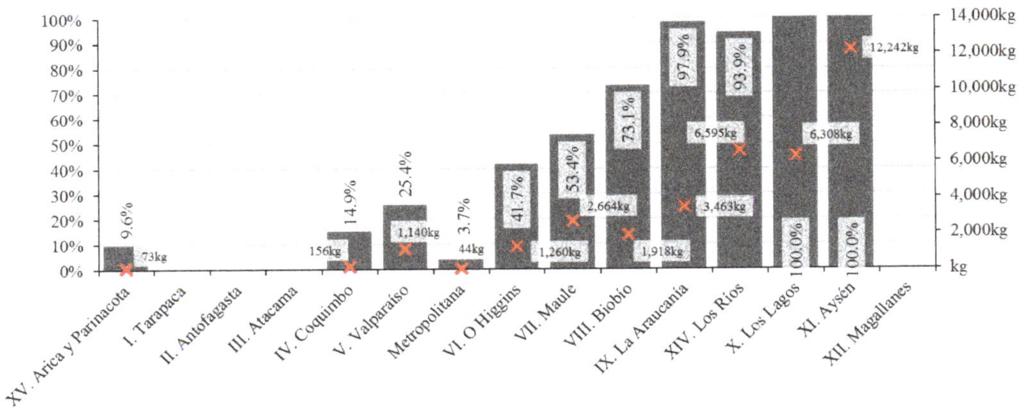

Figure 14. Percentage of dwellings that declare use of firewood for space heating and average annual consumption in kilograms by geographic region [35].

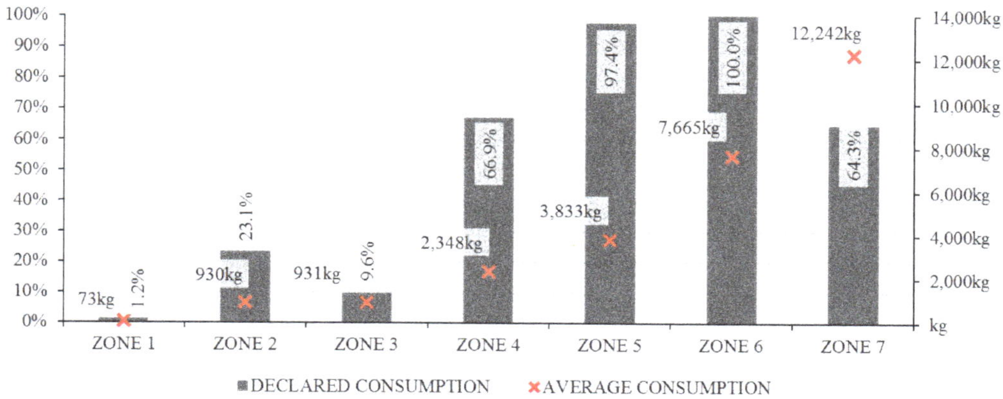

Figure 15. Percentage of dwellings that declare use of firewood for space heating and average annual consumption in kilograms by climatic zone [35].

4.3. Household Consumption Profiles

Pearson Correlation was used to identify existing relationships between the socio-economic capabilities, needs, and space heating behavior of SFHC households. The most significant correlations were then used to identify statistically significant consumption profiles using two-step cluster analysis methods.

4.3.1. Correlation Analysis

The results of the correlation analysis are shown below (Table 2). The largest quantity of correlations occurred among consumption, climatic zone and characteristics of the dwellings, where strong correlations (r = 0.4 to 0.69) were found between dwelling type and materiality of roofs and walls, and between climatic zone, firewood consumption and wall materiality. These correlations are significant as they confirm that any relationship between climatic conditions and space heating behaviour may also be affected by differences in the material characteristics of the dwellings. Then, weak correlations (r = 0.2 to 0.29) were found between climatic zones and dwelling types but also between roof and wall materials in accordance to the different construction techniques used across the country. Interestingly, a very strong (r = 0.741) positive correlation was found between household size and total autonomous income, suggesting that although the SPF system assumes a larger number of dwellers to be a burden, in the SFHC group this may have positive consequences in terms of household income.

When the relationship between household size and income was adjusted by educational level and occupation using Partial Correlation techniques, the *r* values were reduced to 0.715 and 0.534 respectively, suggesting that the occupation of the dweller may play a stronger role in this correlation. Both variables maintained the strongest *r* values when correlated to declared firewood consumption per climatic zone (Table 3). Autonomous income displayed very strong (r = 0.7+) correlations with annual firewood consumption in Zones 1 and 2, and strong (r = 0.4 to 0.69) correlations in Zone 6, while household size showed strong correlations in Zones 1, 2 and 6 and weak correlations in Zone 7. Dwelling type also showed strong and very strong correlations in the same climatic zones, while wall and roof materials displayed only week to moderate correlations.

Table 2. Correlation analysis of selected dwelling characteristics, socio-economic capabilities and household needs [35].

Firewood	*Mat. Roof*	*Mat. Walls*	*Dwe. Type*	*Educ.*	*Occupat.*	*Income*	*h. Size*	*Disability*	*h. Mark*	
0.523	0.143	0.450	−0.299	0.011	0.026	−0.049	−0.092	−0.028	−0.033	*zone*
	−0.122	0.163	−0.074	0.020	0.005	0.052	0.093	0.007	0.038	*firewood*
		0.204	−0.496	−0.005	−0.086	0.104	0.078	0.023	0.018	*material roof*
			−0.452	−0.026	−0.024	−0.006	−0.081	−0.019	−0.012	*material walls*
				−0.009	0.073	−0.076	−0.027	0.046	0.032	*dwelling type*
					−0.077	−0.026	−0.062	−0.048	0.035	*education*
						−0.100	−0.014	−0.028	−0.111	*occupation*
							0.741	0.119	0.147	*a. income*
								0.119	0.147	*household size*
									0.274	*disability*

Interestingly, no significant correlations were found in climatic Zones 4 and 5 although they concentrate a significant percentage of the total firewood consumption nationwide. This may be a result of the ubiquitous use of this resource for space heating purposes in these areas regardless of the socio-economic status of the household and the properties of their dwelling. These results remained consistent when firewood consumption was correlated to household size and autonomous income per geographic region, while stronger correlations in northern and central areas also showed the lowest rates of firewood consumption nationwide (Table 4).

Table 3. Correlations between firewood consumption and selected dwelling and household characteristics by climatic zone [35].

	ZONE 1	ZONE 2	ZONE 3	ZONE 4	ZONE 5	ZONE 6	ZONE 7
material roof	-	-	0.355	-	0.092	−0.293	-
material walls	-	−0.547	0.214	0.021	−0.247	-	−0.243
dwelling type	-	0.702	−0.426	−0.038	0.000	0.510	−0.116
education	−0.006	0.081	−0.085	0.052	0.045	0.021	−0.139
occupation	-	−0.497	0.197	−0.048	−0.190	0.215	0.213
autonomous income	−0.962	0.806	−0.149	0.082	0.097	0.446	0.061
household size	−0.515	0.616	−0.076	0.015	0.184	0.491	0.296
disability	-	0.336	−0.043	0.033	0.027	−0.142	0.164
health mark	−0.061	0.531	0.019	0.046	−0.005	0.057	−0.015

Table 4. Correlations between of annual firewood consumption, household size and autonomous income per geographic region [35].

	XV	I	II	III	IV	V	RM	VI	VII	VIII	IX	XIV	X	XI	XII
household size	−0.515	-	-	-	−0.829	0.443	−0.832	0.240	−0.020	−0.057	0.211	−0.288	0.305	0.296	-
autonomous income	−0.962	-	-	-	−0.784	0.720	−0.654	−0.185	−0.019	−0.058	0.145	−0.202	−0.014	0.061	-

4.3.2. Cluster Analysis

Two-step cluster analysis was used to identify the most defining socio-economic characteristics of the SFHC households in relation to their annual firewood consumption. In terms of number of people per household, two main groups were identified with 3.75 and 7.19 people in average in contrast to three household types identified in the total population (Figure 16).

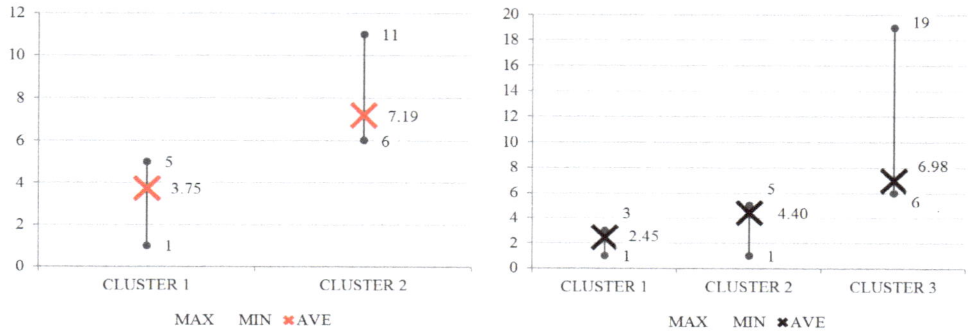

Figure 16. Two-step cluster analysis of the number of users per dwelling of the SFHC group (**left**) and of the total population (**right**) [35].

According to clustering of individuals by age segment, these households can fall into three main categories (Figure 17). The first cluster was the largest and closest to existing definitions of extended families with an average of 5.09 members and a relatively even distribution among child, young, young adults, and adults. This strongly contrasts with the second cluster, which had an average of 2.62 members with a tendency towards elder population and almost null presence of child. This may be related to current ageing tendencies in the Chilean population [39] as well as the effects of regulation prioritizing senior citizens as beneficiaries [33]. The third cluster found might be the closest to traditional definitions of nuclear families having an average of 3.40 members (close to the national average), a marked presence of both adults and young, but an almost null presence of child and elders.

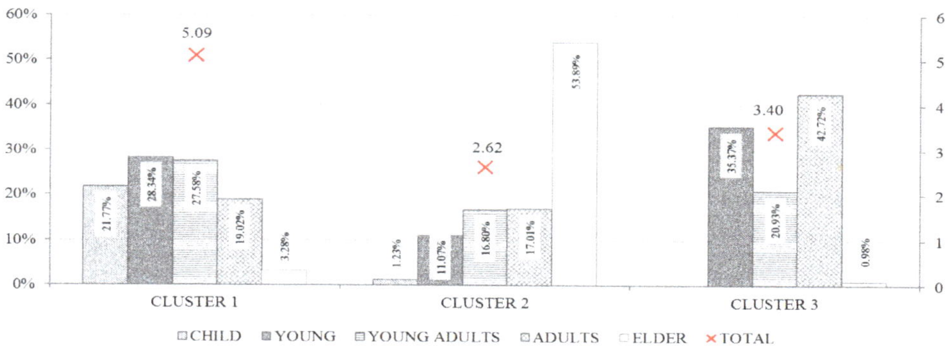

Figure 17. Two-step cluster analysis of the main age segments found in the SFHC target population [35].

In terms of existing relationships between household size, autonomous income, and firewood consumption, two-step clustering identified two main groups, namely the *high consumption* and the *low consumption* household profiles. While low consumption households concentrated both the lowest autonomous income and household sizes, the high consumption households sprawled across the whole range with a higher concentration at the larger household size and income spectrum (Figure 18). The average household size, monthly income, and annual firewood consumption per climatic zone of each household profile are shown below (Table 5). While the average annual use of firewood for space heating in households belonging to the high consumption profile was of 5473 kg with 6.37 users and CL$300,962 of autonomous income, the average annual use in households belonging to the low consumption profile was of 3182 kg with 3.41 users and CL$157,435 of autonomous income.

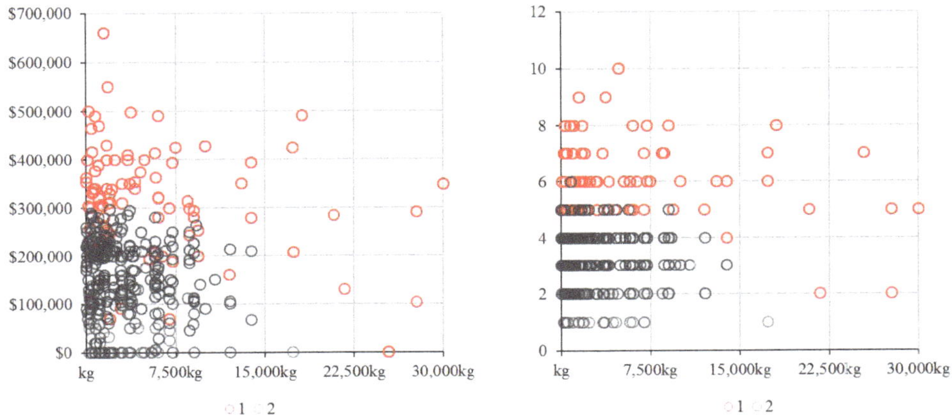

Figure 18. Results of two-step cluster analysis of autonomous income (**left**) and number of users (**right**) by annual firewood consumption (x axis) [35].

Table 5. Average firewood consumption per climatic zone of the main household profiles as found in the SFHC group [35].

		ZONE 1	ZONE 2	ZONE 3	ZONE 4	ZONE 5	ZONE 6	ZONE 7
HIGH CONSUMPTION HOUSEHOLD	*income*	-	$325,000	$236,879	$346,910	$341,120	$263,209	$292,656
	users	-	8.00	5.87	6.41	6.34	5.88	5.69
	consumption	-	1080 kg	1502 kg	2373 kg	4577 kg	7384 kg	15,926 kg
LOW CONSUMPTION HOUSEHOLD	*income*	$191,333	$184,128	$183,241	$168,343	$157,714	$120,641	$130,541
	users	3.40	3.56	3.56	3.58	3.52	3.09	3.12
	consumption	71 kg	316 kg	1248 kg	2277 kg	3854 kg	3949 kg	7448 kg

5. Conclusions

This study used the CASEN 2013 survey dataset to describe the main demographic characteristics of the target users of the SFHC housing program and to identify existing correlations between socio-economic characteristics of different households, their dwellings and space heating behavior. The demographic characteristics were described under four main categories: household size and structure, education and occupation, income and poverty, and disabilities and health issues. Although the SFHC group showed strong similarities to Quintile 1, there were marked differences in terms of household size, overcrowding indexes, presence of internal secondary households and number of nuclei, number of female heads of household, and gender distribution per age segments. Although no significant differences were found in terms of educational level, there were marked contrasts in terms of occupation that may have a direct impact on the energy use of their dwellings. No significant differences were found between the SFHC group and the first quintile in terms of disabilities, poverty, or income, but their access to monetary subsidies and overall health mark were consistently higher suggesting reduced risks to fall into fuel poverty.

This information was analyzed using Pearson correlation and two-step cluster analysis techniques to identify variables that may be used to profile differences in space heating behavior among households. Beyond location-specific variables such as climatic zone, the strongest correlations found suggested that household size and autonomous income may be the best predictors of mean annual firewood consumption. These variables were clustered and two statistically different groups were identified: a high consumption household profile in which a larger number of users and higher autonomous income may result in higher annual firewood consumption, and a low consumption household profile in which a lower income and average number of users may result in lower annual firewood consumption.

The results of this study suggest that it is possible to predict general patterns of operational energy consumption for space heating purposes using discrete socio-economic variables that can be accessible through social housing allocation systems. Although the forecasting capabilities of the identified clusters are not yet conclusive, they provide enough information to foresee general tends among different households before they occupy their dwellings, hence may help to personalize the offering of energy efficiency measures such as, e.g., fact sheets or operation manuals, recommendations for artifacts, furnishings and extensions. Moreover, as these energy consumption profiles can also be assessed before the construction of a dwelling, they may be used in combination to the described demographic characteristics to customize housing solutions and/or to inform decision-making throughout the design process. Although the results of this study focus specifically on firewood consumption under the Chilean social housing policy, similar methods can be applied to explore further environmental variables under similar mass housing policies.

Acknowledgments

The results shown in this paper are part of a Ph.D. currently under development at the Faculty of Architecture, Building and Planning of the University of Melbourne with the financial support of CONICYT, Chile R.E. No. 2033/2012.

Author Contributions

Victor Bunster designed and conducted the research under the supervision of Masa Noguchi. Both authors read and approved the final manuscript.

Conflicts of Interest

The authors declare no conflict of interest.

References

1. United Nations Environment Programme (UNEP). *Global Environment Outlook: Environment for Development*; UNEP: Nairobi, Kenya, 2007.
2. International Energy Agency (IEA). *Energy Efficiency Indicators: Essentials for Policy Making*; IEA: Paris, France, 2014.
3. Zhenjun, M.; Cooper, P.; Daly, D.; Ledo, L. Existing building retrofits: Methodology and state-of-the-art. *Energy Build.* **2012**, *55*, 889–902.
4. Pérez-Lombard, L.; Ortiz, J.; González, R.; Maestre, I.R. A review of benchmarking, rating and labelling concepts within the framework of building energy certification schemes. *Energy Build.* **2009**, *41*, 272–278.
5. Abrahamse, W.; Steg, L.; Vlek, C.; Rothengatter, T. The effect of tailored information, goal setting, and tailored feedback on household energy use, energy-related behaviors, and behavioral antecedents. *J. Environ. Psychol.* **2007**, 27, 265–276.
6. Abrahamse, W.; Steg, L.; Vlek, C.; Rothengatter, T. A review of intervention studies aimed at household energy conservation. *J. Environ. Psychol.* **2005**, *25*, 273–291.
7. Allcott, H. Social norms and energy conservation. *J. Public Econ.* **2011**, *95*, 1082–1095.
8. Ayres, I.; Raseman, S.; Shih, A. *Evidence from Two Large Field Experiments that Peer Comparison Feedback can Reduce Residential Energy Usage*; NBER Working Paper No. 15386; National Bureau of Economic Research (NBER): Cambridge, MA, USA, 2009.
9. Ministerio de Energía. Balance Nacional de Energía, 1993–2012. Available online: http://www.cne.cl (accessed on 25 February 2015). (In Spanish)
10. Castillo, G.; Maldonado, P. *Situación de la energía en Chile: Desafíos para la sustentabilidad*; Programa Chile Sustentable; LOM Ediciones: Santiago, Chile, 2004. (In Spanish)
11. Ministerio de Energía. *Agenda de energía: Un desafío país, progreso para todos*; Ministerio de Energía: Santiago, Chile, 2014. (In Spanish)
12. Corporación de Desarrollo Tecnológico de la Cámara Chilena de la Construcción (CDT-CCHC). *Estudio de usos Finales y curva de oferta de la Conservación de la Energía en el Sector Residencial: Informe Final*; CDT: Santiago, Chile, 2010. (In Spanish)
13. Gómez-Lobo, A. El consumo de leña en el sur de Chile: ¿Por qué nos debe preocupar y qué se puede hacer? *Rev. Ambient. Desarro.* **2005**, *21*, 43–47. (In Spanish)
14. Ministerio de Vivienda y Urbanismo (MINVU). Encuesta Panel de Vivienda 2010: Resultados parciales. Available online: http://www.minvu.cl/ (accessed on 25 February 2015). (In Spanish)
15. Gómez-Lobo, A.; Lima, J.A.; Hill, C.; Meneses, M. *Diagnóstico del Mercado de la Leña en Chile: Informe Final*; Centro de Micro Datos, Departamento de Economía, Universidad de Chile: Santiago, Chile, 2006. (In Spanish)
16. Bustamante, W.; Cepeda, R.; Martínez, P.; Santa María, H. *Eficiencia Energética en Vivienda Social: Un Desafío Posible*; CIP-PUC: Santiago, Chile, 2009. (In Spanish)

17. Celis, F.; Garcia, R.; Trebilcock, M.; Escorcia, O.; Miotto, U.; Díaz, M. Energy analysis of housing in the central-south of Chile. *Arquitecturarevista* **2012**, *8*, 62–75.
18. Ministerio de Vivienda y Urbanismo (MINVU). *Ordenanza General de Urbanismo y Construcciones*; Diario Oficial de la República de Chile: Santiago, Chile, 2014. (In Spanish)
19. Bustamante, W.; Rozas, Y.; Cepeda, R.; Encinas, F.; Martínez, P.; Ubilla, M.; Sierra, A. *Guía de Diseño para la Eficiencia Energética en la Vivienda Social*; Pontifica Universidad Católica de Chile: Santiago, Chile, 2009. (In Spanish)
20. Ministerio de Vivienda y Urbanismo (MINVU). *D.S. No. 255—Reglamenta Programa de Protección del Patrimonio Familiar*; Diario Oficial de la República de Chile: Santiago, Chile, 2006. (In Spanish)
21. Escorcia, O.; García, R.; Trebilcock, M.; Celis, F.; Bruscato, U. Envelope improvements for energy efficiency of homes in the south-central Chile. *Inf. Constr.* **2012**, *64*, 563–574. (In Spanish)
22. García, R.; González, A. Form conditions and energy performance of single-family housing in central south Chile. *Rev. INVI* **2014**, *80*, 111–141. (In Spanish)
23. Celis, F.; Diaz, M.; Echeverria, E.; García, R.; Escorcia, O.; Trebilcock, K. Incidence of architectural configuration on energy efficiency of dwellings in the centre-south of Chile. In Proceedings of the PLEA 2012—28th Conference, Lima, Perú, 7–9 November 2012.
24. Encinas, F.; de Herde, A. Definition of occupant behaviour patterns with respect to ventilation for apartments from the real estate market in Santiago de Chile. *Sustain. Cities Soc.* **2011**, *1*, 38–44.
25. Frederiks, E.R.; Stenner, K.; Hobman, E.V. The socio-demographic and psychological predictors of residential energy consumption: A comprehensive review. *Energies* **2015**, *8*, 573–609.
26. Abrahamse, W.; Steg, L. Factors related to household energy use and intention to reduce it: The role of psychological and socio-demographic variables. *Hum. Ecol. Rev.* **2011**, *18*, 30–40.
27. Guerra Santin, O. Behavioural patterns and user profiles related to energy consumption for heating. *Energy Build.* **2011**, 43, 2662–2672.
28. Ministerio de Vivienda y Urbanismo (MINVU). *D.S. No. 49—Aprueba Reglamento del Programa Fondo Solidario de Elección de Vivienda*; Diario Oficial de la República de Chile: Santiago, Chile, 2011. (In Spanish)
29. Ministerio de Vivienda y Urbanismo (MINVU). *R.E. No. 9020—Deroga Resolución Exenta No. 3577 y Aprueba Itemizado Técnico de Construcción para Proyectos del programa Fondo Solidario de Elección de Vivienda*; Diario Oficial de la República de Chile: Santiago, Chile, 2012. (In Spanish)
30. Ministerio de Vivienda y Urbanismo (MINVU). *R.E. No. 9013—Modifica Resolución Exenta No. 3563 que Aprueba Cuadro Normativo y tabla de Espacios y Usos Mínimos Para el Mobiliario, Para Proyectos del Programa Fondo Solidario de Elección de Vivienda*; Diario Oficial de la República de Chile: Santiago, Chile, 2012. (In Spanish)

31. Ministerio de Desarrollo Social. *D. No. 291—Reglamenta el diseño, uso y aplicación de la Ficha de Protección Social*; Diario Oficial de la República de Chile: Santiago, Chile, 2007. (In Spanish)
32. Ministerio de Desarrollo Social. *R.E. No. 3509—Determina Metodología para Definir Puntaje de Corte de Carencia Habitacional para Calificar al Programa Fondo Solidario de Elección de Vivienda, Regulado por el D.S No. 49*; Diario Oficial de la República de Chile: Santiago, Chile, 2012. (In Spanish)
33. Candia, A.; Castro, R; Codina, G.; Delgado, R; Herrera, R.; Guzmán, J.; Irarrázaval, I.; Larrañaga, O.; Ormazábal, C.; Saracostti, M.; *et al.* Informe final: Comité de expertos Ficha de Protección Social; Comité Protección Social. 2010. Available online: http://www.ministeriodesarrollosocial.gob.cl (accessed on 25 February 2015). (In Spanish)
34. Herrera, R.; Larrañaga, O.; Telias, A. La Ficha de Protección Social. Documento de Trabajo; Santiago, Chile, 2010. Available online: http://www.cl.undp.org/ (accessed on 25 February 2015). (In Spanish)
35. Ministerio de Desarrollo Social. CASEN 2013 Survey Dataset. Available online: http://www.ministeriodesarrollosocial.gob.cl/ (accessed on 25 February 2015).
36. Ministerio de Desarrollo Social. Libro de Códigos Base de Datos Encuesta de Caracterización Socioeconómica Nacional CASEN 2013. Serie Documentos Metodológicos No. 31. Available online: http://www.ministeriodesarrollosocial.gob.cl/ (accessed on 25 February 2015). (In Spanish)
37. Ministerio de Vivienda y Urbanismo (MINVU). *Planos de Zonificación Térmica. Manual de Aplicación Reglamentación Térmica*; MINVU-IC: Santiago, Chile, 2006. (In Spanish)
38. Bacher, J.; Wenzig, K.; Vogler, M. SPSS TwoStep cluster: A first evaluation. *Lehrstuhl Soziol.* **2004**, *2*, 578–588.
39. Instituto Nacional de Estadísticas (INE). Censo 2002: Síntesis de Resultados. Comisión Nacional del VII Censo de Población y Vivienda. Available online: http//:www.ine.cl/ (accessed on 25 February 2015). (In Spanish)
40. George, M.; Graham, C.; Lennard, L. The Energy Penalty: Disabled People and Fuel Poverty. Available online: http://www.le.ac.uk/ (accessed on 9 June 2015).

Analysis of Photovoltaic Applications in Zero Energy Building Cases of IEA SHC/EBC Task 40/Annex 52

Jin-Hee Kim, Ha-Ryeon Kim and Jun-Tae Kim

Abstract: A Net Zero Energy Building (NZEB) considerably reduces the building energy load through high efficiency equipment and passive elements such as building orientation, high insulation, natural daylighting, and ventilation in order to achieve zero energy balance with on-site energy production from renewable energy systems applied to the building. For a Zero Energy Building (ZEB), the heating energy demand can be significantly reduced with high insulation and air tightness, while the cooling energy demand can be curtailed by applying shading device, cross ventilation, *etc.* As such, the electrical energy demand for a ZEB is relatively higher than its heat energy demand. Therefore, the application of a Renewable Energy System (RES) to produce electricity is necessary for a ZEB. In particular, Building Integrated Photovoltaic (BIPV) systems that generate electricity can play an important role for achieving zero energy balance in buildings; BIPVs are multi-functional and there are many ways to apply them into buildings. This study comprehensively analyzes photovoltaic (PV) applications in ZEB cases through the International Energy Agency Solar Heating and Cooling Programme (IEA SHC)/Energy in Buildings and Communities Programme (EBC) Task 40/Annex 52 activities, which include PV installation methods, PV cell type, and electricity generation. The most widely applied RES is the PV system, corresponding to 29 out of a total of 30 cases. Among the roof type PV systems, 71% were non-integrated. In addition, 14 of the 27 cases in which PV systems were applied, satisfied over 100% of the electricity energy demand from the PV system and were found to generate surplus electrical power.

Reprinted from *Sustainability*. Cite as: Kim, J.-H.; Kim, H.-R.; Kim, J.-T. Analysis of Photovoltaic Applications in Zero Energy Building Cases of IEA SHC/EBC Task 40/Annex 52. *Sustainability* **2015**, *7*, 8782-8800.

1. Introduction

The concept of a Zero Energy Building (ZEB) has garnered significant attention as a solution for reducing global greenhouse gas emissions and responding to global warming.

In a report written in 2006 by Torcellini *et al.*, the authors used the following general definition for a ZEB given by The U.S. Department of Energy (DOE) Building Technologies Program: "A net zero energy building (NZEB) is a residential or commercial building with greatly reduced energy needs through efficiency gains such that the balance of energy needs can be supplied with renewable technologies" [1]. The term "net" indicates a balance between energy supplied by the electrical grid system to augment energy produced by a Renewable Energy System (RES), and energy transmitted to the electrical grid system produced as surplus energy by the RES [2]. In 2010, the European Parliament enacted a building energy comprehensive roadmap called the Energy Performance of

Building Directive (EPBD). The EPBD set the goals of a 20% reduction in greenhouse gas emissions and energy consumption by 2020, 20% expansion of renewable energy propagation by 2020, and enforcing application of nearly Zero Energy Building (nZEB) for all new buildings by 2021. nZEB is defined in the EPBD as "The nearly zero or very low amount of energy required should be covered to a very significant extent by energy from renewable sources, including energy from renewable sources produced on-site or nearby" [3].

Through the International Energy Agency (IEA) Solar Heating and Cooling Programme (SHC) Task 13 from June 1989 to June 1994, zero-energy solar house technology development and propagation was jointly pursued by 16 nations over a five-year period [4]. Additionally, 18 countries participated in the IEA SHC/EBC Task 40/Annex 52 from October 2008 to September 2013 on the subject of "Net Zero Energy Solar Buildings" [5], and numerous studies were conducted to establish international standardization of the NZEB definition [6–9].

Voss [10] reported on the status and outlook of low-energy to net zero energy buildings and analyzed the factors that affect energy balance. Studies on various energy calculation methods and load matching for NZEB definition were also performed [11–13]. Optimization tools and methods for NZEB design were comprehensively reviewed [14–17]. Meanwhile, based on conventional international standards and design model cases, the influences of NZEB passive elements and occupant behavior patterns on energy consumption and generation were analyzed [18–20]. In addition, the application sites of renewable energy such as glazed balconies, solar walls, and solar collectors were analyzed through remodeling demonstration buildings evaluated between 1995 to 1998 [21]. In addition, the interaction between the building load matching and grid as a part of Subtask A of the Task 40/Annex 52 was analyzed through various case studies [22]. Among the NZEB cases, a study assessed the performance of the EcoTerra House in Canada, where it was found that its energy consumption was approximately 12.4% of that of regular Canadian homes [23–25]. Research on the NZEB design method and tools, innovative solutions, and energy monitoring methods has been conducted, as well as investigation of the NZEB status of France [26,27]. Through these studies, it was found that the Enerpos building consumed 10 times less energy than the standard building [28]. Heinze [29] monitored settlements in Freiburg, Germany, and analyzed their energy consumption and generation. The heating area and PV installation area, capacity, heating and electricity energy consumption, and total energy demand for each case were shown. In one study, the design process of the Solar XXI building was introduced as a part of the IEA project and the building was reported to have an energy performance 10 times greater than that of a traditional newly constructed office building [30]. Additionally, research on renewable energy systems applied to each case was conducted, and it was concluded that the photovoltaic thermal mechanical ventilation with heat recovery (PV/T MVHR) system is a major means to achieve net zero energy in cold climates [31]. Another study applied a conventional thermal system and a combined heat and power (CHP) integrated biomass system and analyzed to what degree NZEB was satisfied [32]. A study on the PV design method for zero energy was also performed [33,34]. Case study based analysis research has been diversely pursued regarding renewable energy status and building load aspects for each NZEB.

The electricity energy demand for a NZEB can be relatively greater than the thermal energy demand due to the high insulation and high airtight envelope. In this regard, electricity generation via a PV system is essential for a NZEB, and the building integrated photovoltaic (BIPV) system plays an important role for NZEB in terms of functionality and aesthetics. In this study, the renewable energy application status of NZEB cases was investigated and a case study of the PV system application was analyzed. To this end, the "Net Zero Energy Solar Buildings" cases of the joint international project IEA SHC Task 40/Annex 52 were analyzed and, among them, the cases with a PV system applied were analyzed regarding the PV installation type, cell type, and PV power production. Additionally, the analysis results were used to perform a comparative analysis of the electricity energy demand and generation for PV-applied NZEB.

2. IEA SHC/EBC Task 40/Annex 52 Database

IEA SHC/EBC Task 40/Annex 52 is composed of three subtasks. Subtask A develops the international standard of the existing NZEB definition and concept along with monitoring and guide for NZEB verification. Subtask B pertains to the NZEB design tools and simulation, providing the design priorities, detailed design rules, and guide for NZEB. Subtask C concerns advanced building design and technology for the purpose of development and testing of technologies that provide the foundation for construction and demonstration projects and joint international projects. In addition, related studies have been conducted including the construction of an NZEB database.

In this study, 30 cases of the IEA SHC/EBC Task 40/Annex 52 were categorized by country, building use, climate, and renewable energy application. Table 1 [6] shows an overview of NZEB cases and these were cases selected based on the following criteria:

(1) Innovative solution sets and/or innovative technologies clearly identified for each project (ventilation/daylighting/architectural integration);
(2) Net ZEB Lessons learned (feedback from architect/builder);
(3) Net ZEB Energy performance <50% standard buildings (primary and final energy);
(4) Energy supply/integration of renewable energy;
(5) Monitoring mandatory (energy measurements);
(6) Mismatch management (nearly or net ZEB);
(7) Indoor environment data (temperature, humidity, illuminance, *etc.*).

From Table 1, the NZEB cases can be categorized into residential (40%) and non-residential (60%), which includes office and educational buildings. Categorization by country showed that the NZEB database number was greatest for France, and other European nations including Germany and Austria participated in the NZEB project for database construction.

The NZEB cases showed differing heating and cooling load properties depending on the climate of the region, and there were three types of load distribution according to the climate characteristics. Other than the cooling load being dominant in tropical climate regions, the northwestern Europe regions having a west coast oceanic climate and the southern Europe regions having a Mediterranean climate, were all dominated by heating and cooling loads while the central and eastern Europe regions having a continental climate were dominated by heating load.

Overall, there were 30 NZEB cases; more than one renewable energy system were installed for some buildings. Renewable energy systems were complexly applied for all cases except three, among others; the cases using solar energy were the most. Of all 30 cases, a PV system accounted for 97% (29 cases), solar thermal (ST) accounted for 63% (19 cases), geothermal (GT) accounted for 33% (10 cases), and wind and biomass systems accounted for less than 20%. Thus, PV systems were the most widely applied.

Table 1. Overview of net zero energy building (NZEB) cases for International Energy Agency Solar Heating and Cooling Programme (IEA SHC)/Energy in Buildings and Communities Programme (EBC) Task 40/Annex 52.

Building Use	Building Name	Location	Climate	Renewable Energy					
				PV	ST	Wind	GT	Biomass	Biomass CHP
Residential	Casa Zero Energy House	Italy	HCD	√	√		√		
	Kraftwerk B	Switzerland	HD	√	√			√	
	Energy Flex House	Denmark	HCD	√	√		√		
	Single Family Building(Riehen)	Switzerland	HD	√	√		√		
	Riverdale	Canada	HD	√	√				
	Solarsiedlung am Schlierberg	Germany	HD	√					√
	"Le Charpak", IESC Cargese	France	HCD	√	√	√			
	Plus Energy Houses Weiz	Austria	HD	√					
	EcoTerra House	Canada	HD	√	√		√		
	LIMA	Spain	HCD	√	√				
	Leaf House	Italy	HCD	√	√		√		
	Kleehäuser	Germany	HD	√	√	√			√
Office	SOLAR XXI	Portugal	HCD	√	√				
	Marche Kempthal	Switzerland	HD	√			√		
	Elithis Tower	France	HCD	√				√	
	Ilet du Centre	Reunion	CD	√					
	Green Office	France	HCD	√					
	Pixel Building	Australia	HCD	√		√		√	
	Meridian Building	New Zealand	HCD		√	√			
	CIRCE Zaragoza	Spain	HCD	√	√	√	√	√	
	Villach Offices & Apartment	Austria	HD	√	√				√

Table 1. *Cont.*

Building Use	Building Name	Location	Climate	Renewable Energy					
				PV	ST	Wind	GT	Biomass	Biomass CHP
Educational	Primary School of Laion	Italy	HCD	√	√		√		
	ENERPOS	Reunion	CD	√					
	Pantin Primary School	France	HCD	√	√				
	Limeil Brevannes School	France	HCD	√	√		√		
	ZEB@BCA Academy	Singapore	CD	√					
	Day Care Centre	Germany	HD	√	√		√		
	Primary School Hohen Neuendorf	Germany	HD	√					√
	LYCEE KYOTO HIGH SCHOOL	France	HCD	√					√
Other	Alpine Refuge-Schiestlhaus	Austria	HD	√	√				

Pertaining to a NZEB, it is essential to balance (zero) import and export over a time period. In some cases embodied energy or embodied emissions in materials also have to be balanced off. Figure 1 graphically depicts the balance of import (delivered energy) against export (feed-in energy) on the *x* and *y* axes, respectively.

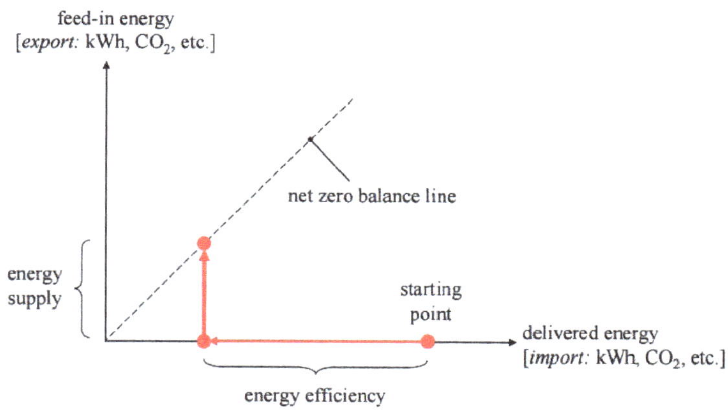

Figure 1. Graph representing the net zero balance of a Net ZEB [22,35].

The starting point may represent the performance of a new building built according to the minimum requirements of the building code or the performance of an existing building prior to renovation work. The general pathway to achieve an NZEB basically consists of two steps, namely reducing the energy demand (*x*-axis) by means of energy efficiency measures, and generating electricity, or other energy carriers, by means of energy supply options to obtain enough credits (*y*-axis) to achieve the balance [35].

Figure 2 shows the relationship between the final energy demand and renewable energy supply of the NZEB cases according to the building use and climate. According to the building use, buildings were classified with symbols such as residential, office, and education, among others. In addition, according to the climate, the buildings were classified as Heating and Cooling Dominated (HCD: Yellow), Heating Dominated (HD: Red), and Cooling Dominated (CD: Blue).

Based on the analysis results, it was observed that 80% out of 30 cases utilized renewable energy sources to either achieve zero energy or generate additional energy. In particular, all residential buildings achieved ZEB. In office building cases, five cases (55%) met the ZEB range of building energy demand and RES were more widely used than in other buildings. Educational buildings totaled six cases (75%) and RES production was under 80 kWh/m^2·year. With most of the buildings, the total energy demand was lower than that of the other buildings because of the differences in schedule. With climate criteria, all CD buildings achieved ZEB. HD buildings were 10 cases (91%) and HCD buildings were 11 cases (69%). The achievement rate of the buildings that required heating and cooling thus was relatively low. Additionally, it was found that some cases did not achieve zero energy due to an insufficient renewable energy supply compared to the energy demand. This was determined to be due to malfunctioning of the applied renewable energy system [36].

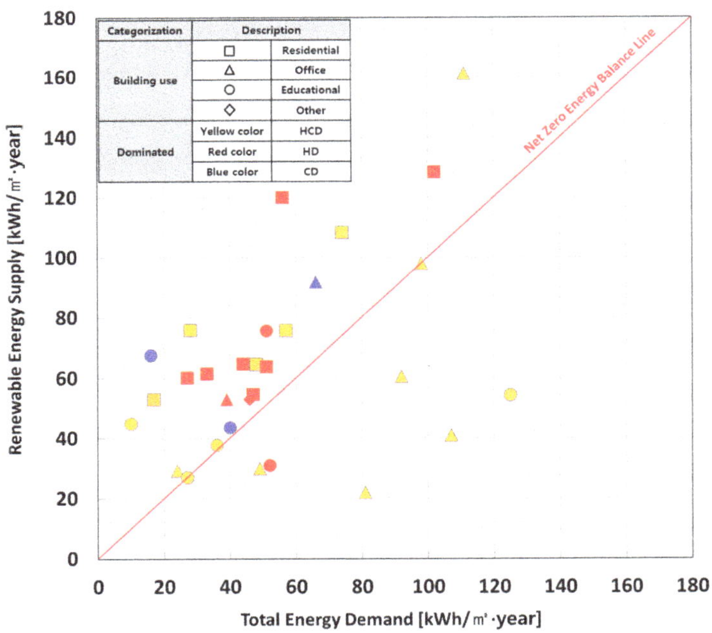

Figure 2. Net zero energy balance status.

3. PV Application Status

Regarding renewable energy systems for NZEB, a number of aspects should be considered before application to a building such as thoroughly conducting an assessment of the foundation from the initial stages of construction due to the ground heat exchanger being buried for geothermal systems and maintaining a sufficient safety distance for the wind effect and blades for wind power systems. On the other hand, solar energy systems have the advantage of convenient installation on existing buildings as well as new buildings compared to other renewable energy sources. In particular, PV and BIPV systems are important renewable energy facilities for NZEB in terms of satisfying the electricity energy demand of buildings. In this section, analyses of PV applied NZEB cases were conducted regarding the PV installation type, cell type, and amount of PV power production. In addition, a comparison of the electricity energy demand and amount generated was performed.

3.1. NZEB Electricity Energy Demand

Figure 3 shows the electricity energy demand and total energy demand for each building purpose of the NZEB cases. The average electricity energy demand was higher than the average total energy demand. In office buildings, the average total energy demand and average electrical energy demand were 69 kWh/m^2 year and 55 kWh/m^2·year, respectively. The electrical energy demand therefore covered 80% of all energy demand, and that of other buildings. Thus, the NZEB cases had high electricity energy demand in comparison to the total energy demand.

The total energy demand was 2–125 kWh/m^2·year and the deviation was found to greatly differ depending on the building purpose. The deviation for educational buildings is notable for being large, and this was ascribed to the different schedules of elementary and high schools despite both being educational buildings. The low electricity energy demand and total energy demand of educational buildings compared to other purpose buildings result from shorter occupancy time and inclusion of vacation periods. The electricity energy demand was determined to be on average 29–55 kWh/m^2·year and the electricity energy demand was found to be highest for offices. Based on the analysis results, residential and educational buildings had lower electricity energy demand compared to office buildings because residential and educational buildings used natural gas or biofuel in place of electricity energy such as an Electric Heat Pump (EHP) for heating and cooling. Seventeen cases with 100% electricity energy demand for the total energy demand were found to use an EHP for heating and cooling. Electrical energy demand was 100% in 17 cases (57%) among all 30 buildings and these cases used an EHP for heating and cooling. The rest of the NZEB cases fulfill an estimated 50% of their energy demands with electricity.

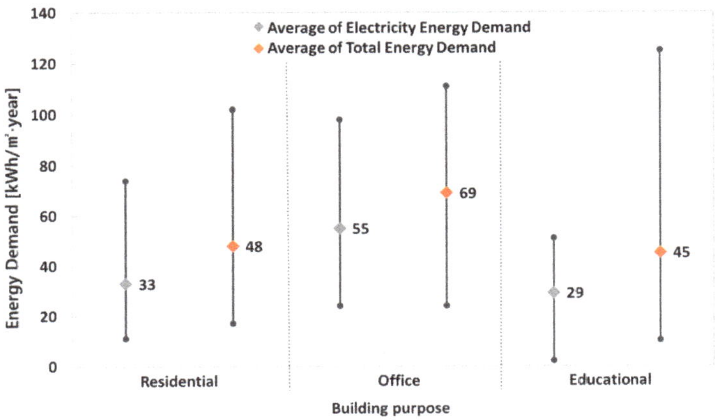

Figure 3. Average energy demand for building use.

3.2. Installation Type Analysis

PV systems can be grouped into many categories such as skyward element, elevation elements, and so on. PV application methods are also variable. Above all, a PV system applied on a roof can be installed independently and at an inclined angle favorable to collect solar radiation. If a PV system with various colors, sizes and shapes were applied on the façade, buildings could be improved aesthetically. Moreover, in the cases of PV installed as shading devices, they could be used as design elements as well as for light control and electricity production. Categorization was conducted according to the installation methods, namely: the non-integrated types, where the PV module is installed by adding a holder or structure on a flat roof or a pitched roof, and the integrated types, where the PV module forms part of the building envelope exterior (see Figures 4 and 5). In this section, the PV systems applied to NZEB are categorized according to the installation type and building factors of the roof and facade.

Figure 4. Categorization of PV installation method: integrated type [33].

Figure 5. Categorization of PV installation method: non-integrated type [33].

The 41 PV systems from the 29 NZEB cases were classified by installation type. Figure 6 shows the PV installation type as building elements by building use. It was analyzed that the 28 PV systems among the 41 PV systems were installed on the roofs. In particular, in the cases of residential building, most of PV systems were applied as building element with roof. For office and educational buildings, they were secondly applied as a façade (Figure 6). In the case of roofs, 29% of the cases employed the integrated method for complete integration with the building while the non-integrated method was applied in 71% of the cases. Additionally, the integrated type was mostly applied for residential buildings. In the case of façades, PV systems were applied to the façade as not only the building envelope but also on the balcony and to provide shade. Furthermore, more cases were found in educational buildings because of the larger effective area for PV installation than residential buildings. PV modules were dominantly applied in the form of the backsheet type while the Glass to Glass PV was applied to the roof and facade with shade functionality to control daylight intake.

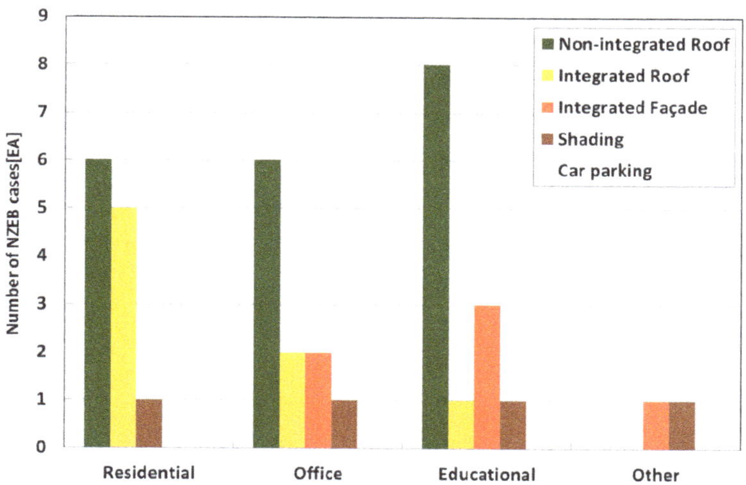

Figure 6. PV application for the NZEB cases.

Among the integrated type cases, building integrated photovoltaic-thermal (BIPVT) was also optionally installed to reduce the building heating load by supplying PV backside heat into the building as well as electricity generation. An amorphous silica applied BIPVT was installed as an integral system on the 30° pitched roof of the EcoTerra House, which was a case that considered BIPVT from the early stages of construction. Figure 7 presents the BIPVT roof and conceptual diagram applied to EcoTerra House. The gap between the inlet air temperature and the outlet was more than 40 °C. The pre-heat source was used for heating and hot-water supply [37].

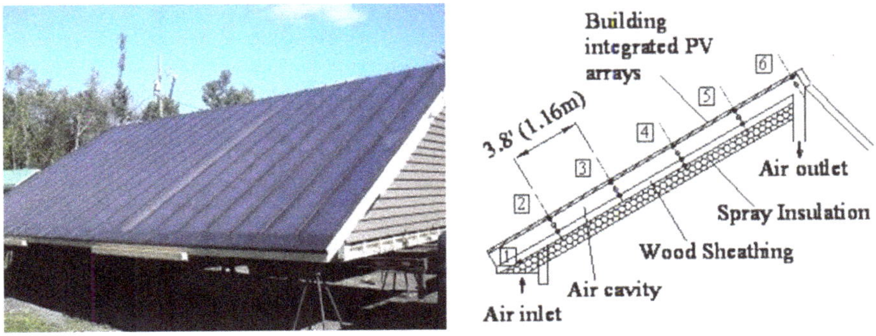

Figure 7. BIPVT case: EcoTerra House [22,38].

For the Solar XXI, BIPVT with poly-crystalline PV modules was installed in the façade in an integral form such that an air cavity was created between the PV backside and the building envelope. The system thus constitutes an outer layer (PV panels) and an inner layer (building envelope). The air gap formed between these two layers interacts with the indoor environment by means of two vents (dampers), one located in the upper part and the other located in the lower part of the brick wall. In addition, the system was designed to allow ventilation of the gap exclusively with outdoor air. The gap ventilation is simply imposed by the occupants who may maneuver the vents manually in the desired position according to their individual comfort needs and weather conditions (Figure 8) [39].

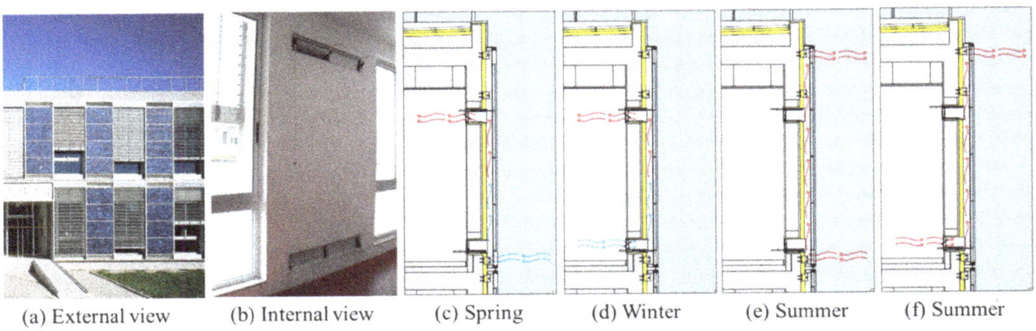

Figure 8. BIPVT case: Solar XXI [39]: seasonal operation mode.

3.3. Cell Type Analysis

The crystalline-Silicon (c-Si) and thin film modules were applied for the PV systems in the NZEB. In general, thin film modules have lower efficiency than c-Si. On this basis, the PV power production of the thin film modules also would be lower. PV power production per unit area by cell type was analyzed in this section. The cases with various cell types applied simultaneously to a building were considered as well. 43 PV systems from the 29 NZEBs were classified by cell type. The cell types of PV systems applied to NZEB were categorized as shown in Figure 9, and analyzed to be more than 77% mono-crystalline and poly-crystalline silicon. Figure 10 shows the installation method for each cell type depending on the building integration. The non-integrated installation was on the roof and façade, whereas the integrated installation on the roof and façade was not integrated with the building but performed additional functions of shading. To analyze the installation method based on the cell type, the parking cases were not considered in the analysis, as installation was not done on the building. Analysis results revealed that most of the amorphous cell cases employed an integrated application. In the case of the crystalline cell type, non-integrated application was employed in comparison to the other cell types. This was ascribed to reduced efficiency by the temperature increase of the crystalline cell.

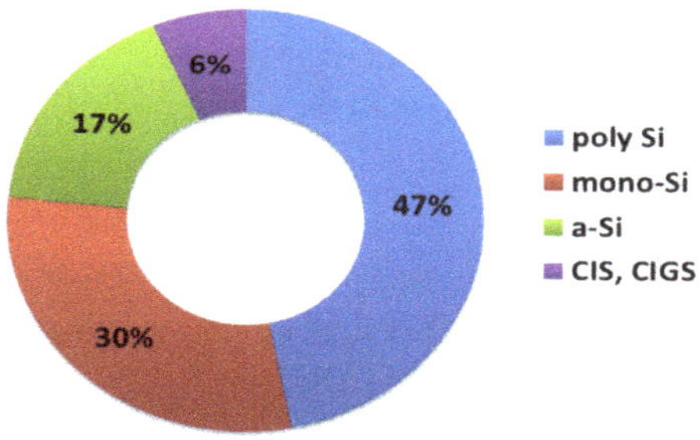

Figure 9. Categorization of cell type.

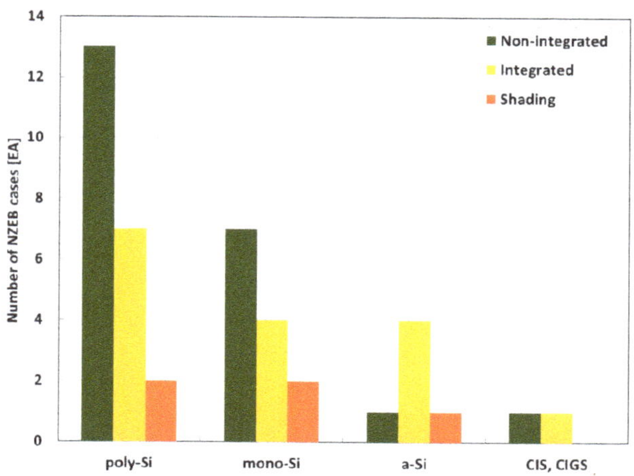

Figure 10. Installation methods depending on building integration

3.4. PV Power Generation

The power generation amount of a PV system is dependent on climate conditions, which determines the amount of radiation in the given region as well as the slope and orientation. It is also influenced by the power generation characteristics of cells and installation conditions. As such, comparing the power generation performance of the NZEB cases is difficult due to differences in region and installation conditions. This study investigated the amount of power generation of PV systems installed in NZEB.

Figure 11 shows the nominal power of each PV system and the annual power generation for 27 NZEB cases arranged in ascending order.

From the graph, it can be deduced that case F has a larger amount of power generation than case E even though the two cases have the same nominal power. The measured generation of case F is much greater *versus* its nominal power. This can be explained by case F incorporating a tracking PV system, unlike other cases.

Case D has a smaller nominal power compared to case E, but a higher measured generation. Case H has a larger nominal power compared to both cases F and G but a lower measured generation. Cases D, F, J, L, M, O, and Q showed outstanding power generation *versus* nominal power, whereas cases B, H, R, S, U, and AA showed poor generation. Compared to case Z, case AA had a far larger nominal power but only a slightly higher measured generation, indicating low efficiency. While other cases involved installation on roofs or walls at an angle between 10 to 50 degrees, part of the PV system of case AA (approx. 52 kW_P) was installed perpendicularly on a wall at 90 degrees.

Six cases featured PV systems installed on walls, but only four cases involved perpendicular wall surfaces (90 degree angle). For all four cases, the systems were partially installed at 90 degrees. Even for cases with PV systems installed on walls, the installation angle was set to maximize radiation. Among all 27 cases, case F (Pixel building, Australia) and case M (Ilet du Centre, Reunion) had the most outstanding power generation *versus* nominal power, while case H (Energy Flex House, Denmark) was the poorest.

Figure 12 shows a graph of the annual PV power generation per unit area depending on the electricity energy demand of the 27 NZEB cases with PV applied. From the figure, the respective plots are distributed away from the PV generation/Electricity demand balance line. 15 cases (56%) had 100% electricity energy demand for the building in 27 cases. Furthermore, power generation in three cases with a CD climate offset the total energy demand. In addition, 14 cases (52%) of the subject buildings satisfied over 100% of the electricity energy demand from the PV system and were found to generate surplus electrical power. However, some cases showed very low PV power generation compared to the electricity energy demand. These cases were determined to have low PV power generation due to a malfunctioning system or because a low efficiency PV module was used [37]. Wind power was also installed for three cases (depicted by green colored symbols in Figure 12) to generate electrical power in addition to the PV system. For one case, the building satisfied the electricity demand, even when only a PV system was installed. Thus wind power generated surplus energy. Although the PV system was augmented with wind power in the second case, the building could not satisfy the electricity demand. Considering the third case, the building could not satisfy the electricity demand until wind power was integrated with the PV system.

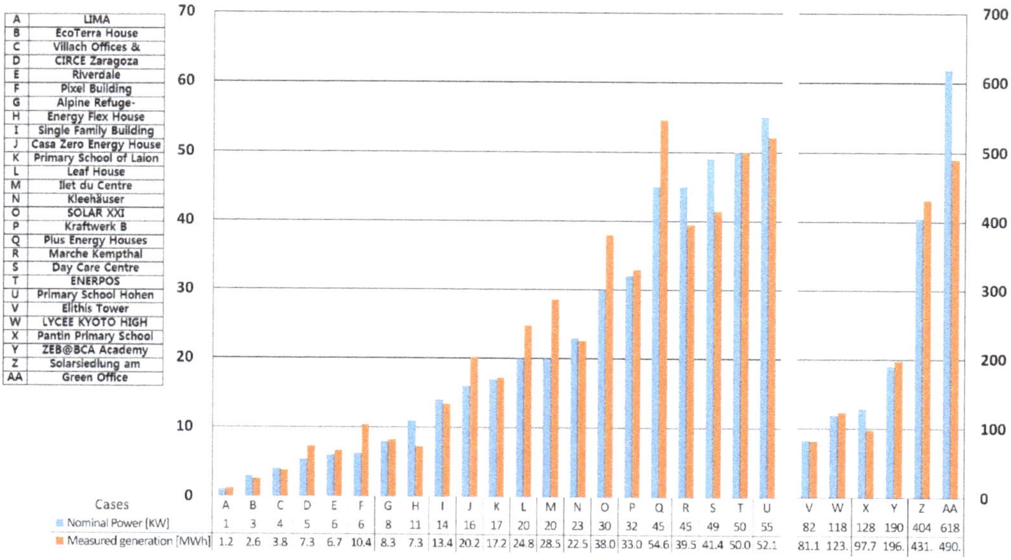

Figure 11. Nominal power and measured generation of PV system in NZEB cases.

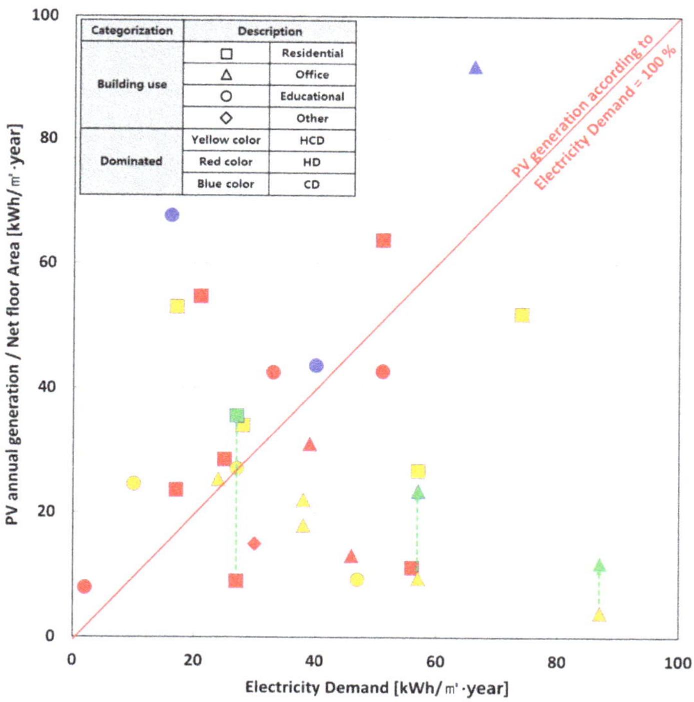

Figure 12. Balance of PV power generation and electricity energy demand of NZEB cases.

4. Conclusions

In this research, through a case study of IEA SHC/EBC Task 40/Annex 52 NZEB, energy aspects of the PV system such as application methods and energy balance with a consideration of demand and production were analyzed. Results of the analysis showed that 70% of all the 41 types of PV systems that were applied to 30 cases of NZEB were installed on areas of roofs with effective solar energy gain; 34% of the total number of PV systems were applied as fully integrated BIPV systems. The remaining PV systems were attached as non-integrated PV systems on building facades or roofs. Therefore, it can be seen that there were not many cases of fully integrated PV systems that replaced building envelope materials. It can be said that fully integrated BIPV systems, when applied as building components, have great impacts on buildings in terms of architectural appearance and energy performance, as well as having effects on the PV system's energy production. However, there are many aspects to consider when applying such systems to buildings, including economic viability, the harmonization of PV installation with the building exterior and structure, and surrounding environment; hence, the application of BIPV systems in buildings is difficult. Therefore, a variety of solutions, and consistent effort, will be required to make BIPV systems in NZEB a reality.

On the other hand, additional research results have showed that 57% of the NZEB cases use electricity for 100% of the energy demands for heating, cooling, and DHW. The rest of the NZEB cases fulfill an estimated 50% of their energy demands with electricity. Therefore it can be concluded that the electricity generation of PV systems in NZEB is essential.

Results indicated that for PV systems in NZEB cases, about 20% of the cases had relatively poor PV generation performance despite having similar levels of nominal power. For improved PV generation in NZEB, various attempts and efforts must be made to optimize running and design of PV systems including solar radiation and temperature characteristics. For those buildings that use electricity as their entire energy source, zero energy balance can be achieved through the installation of PV systems. Additionally, in 14 of the 27 cases in which PV systems were applied, all the electricity required in the buildings was supplied through electricity production by the PV system.

It is concluded that PV systems significantly contribute to net zero energy status for cases that demand electricity as their main energy source. However, in some cases, a wind power system was used to help achieve net zero energy balance, such that further analysis will be required to determine the effects of the combined application of PV systems with other renewable energy systems, such as wind power, solar thermal and geothermal energy systems. Furthermore, it is also important to analyze the characteristics of BIPV systems in relation to grid interaction and energy load matching.

Acknowledgments

This work was supported by the New & Renewable Energy Core Technology Program (No. 2012T100100065) and the Human Resources Program in Energy Technology (No. 20134010200540) of the Korea Institute of Energy Technology Evaluation and Planning (KETEP) granted financial resource from the Ministry of Trade, Industry & Energy, Republic of Korea.

Author Contributions

Each author contributed equally to this work. Jun-Tee Kim designed the research concept and did critical revision and final approval of the manuscript to be published. Ha-Ryeon Kim developed the introduction section and approached to zero energy building and PV system. Data acquisition, analysis and interpretation and drafting of manuscript were made by Jin-Hee Kim. All authors read and approved the manuscript.

Conflicts of Interest

The authors declare no conflict of interest.

List of Abbreviations

1. NZEB	19. net zero energy building
2. nZEB	20. nearly zero energy building
3. ZEB	21. zero energy building
4. RES	22. renewable energy system
5. IEA	23. International Energy Agency
6. SHC	24. Solar Heating and Cooling Programme
7. EBC	25. Energy in Buildings and Communities Programme
8. EPBD	26. Energy Performance of Buildings Directive
9. PV	27. photovoltaic
10. BIPV	28. building integrated photovoltaic
11. BAPV	29. building attached photovoltaic
12. BIPVT	30. building integrated photovoltaic-thermal
13. HCD	31. heating and cooling dominated
14. HD	32. heating dominated
15. CD	33. cooling dominated
16. PV/T MVHR	34. photovoltaic thermal mechanical ventilation with heat recovery
17. CHP	35. combined heat and power
18. EHP	36. electric heat pump

References

1. Marszal, A.J.; Heiselberg, P.K. *A Literature Review of Zero Energy Buildings (ZEB) Definitions*; Aalborg University: Copenhagen, Denmark, 2009.
2. Sartori, I.; Napolitano, A.; Voss, K. Net zero energy buildings: A consistent definition framework. *Energy Build.* **2012**, *48*, 220–232.
3. Eduardo, M.; Peter, W.; Marianna, P. *Implementing the Energy Performance of Buildings Directive*; EPBD: Brussels, Belgium, 2013.

4. Hestnes, A.G. *Task 13: Advanced Solar Low-Energy Building Final Report*; International Energy Agency: Paris, France, 1996.
5. Net Zero Energy Solar Buildings. Available online: http://task40.iea-shc.org/ (accessed on 3 July 2014).
6. Grade, F.; Donn, M. Solution Sets and Net Zero Energy Buildings: A Review of 30 Net ZEBs case studies worldwide. Available online: https://www.iea-shc.org/data/sites/1/publications/T40A52-DC-TR1-30-Net-ZEBs.pdf (accessed on 3 July 2015).
7. Voss, K.; Sartori, I.; Lollini, R. Nearly-zero, net zero and plus energy buildings-how definitions & regulations affect the solutions. *REHVA J.* **2012**, *49*, 23–27.
8. Mlecnik, E.; Attia, S.; Loon, S.V. Net zero energy building: A review of current definitions and definition development in Belgium. *Passiv. House Symp.* **2011**, *10*, 250–262.
9. Mlecnik, E. Defining nearly zero-energy housing in Belgium and the Netherlands. *Energy Effic.* **2012**, *5*, 411–431.
10. Voss, K.; Musall, E.; Lichtme, M. From low-energy to net zero energy buildings: Status and perspectives. *J. Green Build.* **2011**, *6*, 46–57.
11. Noris, F.; Musall, E.; Salom, J.; Berggren, B.; Jensen, S.Ø.; Lindberg, K.; Sartori, I. Implications of weighting factors on technology preference in net zero energy buildings. *Energy Build.* **2014**, *82*, 250–262.
12. Marszala, A.J.; Heiselberga, P.; Bourrelleb, J.S.; Musallc, E.; Voss, K.; Sartori, I.; Napolitanoe, D.A. Zero energy building—A review of definitions and calculation methodologies. *Energy Build.* **2011**, *43*, 971–979.
13. Salom, J.; Widen, J.; Candanedo, J.; Sartori, I.; Voss, K.; Marszal, A. Understanding net zero energy buildings: Evaluation of load matching and grid interaction indicators. In Proceedings of Building Simulation 2011: 12th Conference of International Building Performance Simulation Association, Sydney, Australia, 14–16 November 2011; pp. 2514–2521.
14. Belleri, A.; Noris, F.; Oberegger, U.F.; Lollini, R. Evaluation tool for net zero energy buildings: Application on office building. In proceedings of the World Sustainable Energy Days, Wels, Austria, 27 February–1 March 2013.
15. Attia, S.; Hamdy, M.; O'Brien, W.; Carlucci, S. Computational optimization for zero energy buildings design: Interviews results with twenty eight international experts. In Proceedings of the 13th Conference of International Building Performance Simulation Association, Chambery, France, 26–28 August 2013; pp. 2187–2194.
16. Attia, S.; Hamdy, M.; O'Brien, W.; Carlucci, S. Assessing gaps and needs for integrating building performance optimization tools in net zero energy buildings design. *Energy Build.* **2013**, *60*, 110–124.
17. Carlucci, S.; Pagliano, L. An optimization procedure based on thermal discomfort, minimization to support the design of comfortable net zero energy buildings. In Proceedings of the 13th Conference of International Building Performance Simulation Association, Chambery, France, 26–28 August 2013; pp.3690–3697.

18. Sartori, I.; Candanedo, J.; Geier, S.; Lollini, R.; Garde, F.; Afhienitis, A.; Pagliano, L. Comfort and energy efficiency recommendations for net zero energy buildings. In Proceedings of the Eurosun-International Conference on Solar Heating, Cooling and Buildings, Graz, Austria, 28 September–1 October 2010.
19. Bucking, S.; Fhienitis, A.; Zmeureanu, R. Optimization of net-zero energy solar communities: Effect of uncertainty due to occupant factors. In Proceeding of the 30th ISES Biennial Solar World Congress, 28 August–2 September 2011; pp. 1523–1528.
20. Wagner, A.; Lutzkendorf, T.; Voss, K.; Spars, G.; Maas, A.; Herkel, S. Performance analysis of commercial buildings-results and experience from the German demonstration program "Energy Optimized Building(EOB)". *Energy Build.* **2014**, *68*, 634–638.
21. Voss, K. Solar energy in building renovation-results and experience of international demonstration buildings. *Energy Build.* **2000**, *32*, 291–302.
22. Voss, K.; Sartori, I.; Napolitano, A.; Geier, S.; Gonzalves, H.; Hall, M.; Heiselberg, P.; Jose, W.; Candanedo, A.G.; Musall, E.; *et al.* Load matching and grid interaction of net zero energy buildings. In Proceedings of the EuroSun 2010: International Conference on Solar Heating, Cooling and Buildings, Graz, Austria, 28 September–1 October 2010; pp. 28–35.
23. Dorion, M.; O'Brien, W.; Athienitis, A. Energy performance, comfort and lessons learned from a near net zero energy solar house. *ASHRAE Trans.* **2011**, *117*, 585–604.
24. Bucking, S.; Athienitis, A.; Zmeureanu, R.; O'Brien, W.; Doiron, M. Design optimization methodology for a near net zero energy demonstration home. In Proceedings of the EuroSun 2010: International Conference on Solar Heating, Cooling and Buildings, Graz, Austria, 28 September–1 October 2010.
25. O'Brien, W.; Athienitis, A.; Bucking, S.; Doiron, M.; Kesik, T. A study of design tools and processes through a near net-zero energy house redesign. In Proceedings of the EuroSun 2010: International Conference on Solar Heating, Cooling and Buildings, Graz, Austria, 28 September–1 October 2010.
26. Lenoir, A.; Wurtz, E.; Garde, F. Zero energy buildings in France: Overview and feedback. *ASHRAE Trans.* **2011**, *117*, 817–832.
27. Lenoir, A.; Garde, F.; Ottenwelter, E.; Bornarel, A.; Wurtz, E. Net Zero Energy Building in France: From Design Studies to Energy Monitoring. Available online: http://www.iea-shc.org/data/sites/1/publications/Task40c-Net_Zero_Energy_Buildings_in_France_From_Design_Studies_to_Energy_Monitoring.pdf (accessed on 3 July 2014).
28. Franco, M.; Baird, G.; Gar de, F.; Lenoir, A. Environmental Design and Performance of the ENERPOS Building, Reunion Island, France. Available online: http://task40.iea-shc.org/data/sites/1/publications/DC-TP12-Garde-2011-06.pdf (accessed on 4 July 2014).
29. Heinze, M.; Voss, K. Goal: Zero energy building exemplary experience based on the solar estate *Solarsiedlung Freiburg am Schlierberg,* Germany. *J. Green Build. Fall* **2009**, *4*, 93–100.

30. Aelenei, L.; Goncalves, H.; Rodrigues, C. The road towards "Zero Energy" in buildings: Lessons learned from SOLARXXI building in Portugal. In Proceedings of the EuroSun 2010: International Conference on Solar Heating, Cooling and Buildings, Graz, Austria, 28 September–1 October 2010; p. 8.
31. Noguchi, M.; Higuchi, Y.; Udagawa, M.; Kim, J.T.; Kirkwood, B. Calibrating the impact of a photovoltaic thermal mechanical ventilation heat recovery system on the delivery of net zero-energy housing in Scotland. In Proceedings of the 6th Energy Forum on Solar building Skins, Bressanone, Italy, 6–7 December, 2011; pp. 113–117.
32. Mohamed, A.; Hasan, A.; Siren, K. Fulfillment of net-zero energy building (NZEB) with four metrics in a single family house with different heating alternatives. *Appl. Energy* **2014**, *114*, 385–399.
33. Scognamiglio, A.; Garde, F. Photovoltaics architectural and landscape design options for Net Zero Energy Buildings, towards Net Zero Energy Communities: Spatial features and outdoor thermal comfort related considerations. *Prog. Photovolt. Res. Appl.* **2014**, doi:10.1002/pip.2563.
34. Scognamiglio, A.; Rostvik, H.N. Photovoltaics and zero energy buildings: A new opportunity and challenge for design. *Prog. Photovolt. Res. Appl.* **2013**, *21*, 1319–1336.
35. Sartori, I.; Napolitano, A.; Marszal, A.; Pless, S.; Torcellini, P.; Voss, K. Criteria for definition of net zero energy buildings. In Proceedings of the EuroSun 2010: International Conference on Solar Heating, Cooling and Buildings, Graz, Austria, 28 September–1 October 2010.
36. Conseil Regional Poitou. *Technip, Suivi des Consommations September 2011 Rapport Final*; Conseil Regional Poitou: Poitiers, France, 2011; p. 13. (In French)
37. Doiron, M.A. Whole-building energy analysis and lessons learned for a near net-zero energy solar house. Master's Thesis, A thesis in the department of building, Civil and Environmental Engineering, Concordia University, Montreal, QC, Canada, February 2011.
38. Carsten, W. Equilibrium^{TM} house with ground-coupled heat pump and BIPV/T-solar system (Ecoterra^{TM}home). Available online: http://www.annex32.net/pdf/System_concepts/System_Concept_CA_EcoTerra_NZEB.pdf (accessed on 30 March 2015).
39. Laura, A.; Helder, G.; Carlos, R. The Road towards "Zero Energy" in Buildings: Lessons Learned from SOLARXXI Building in Portugal. Available online: http://task40.iea-shc.org/data/sites/1/publications/Task40c-The_Road_Towards_Zero_Energy_in_Buildings.pdf (accessed on 4 July 2014).

MDPI AG
Klybeckstrasse 64
4057 Basel, Switzerland
Tel. +41 61 683 77 34
Fax +41 61 302 89 18
http://www.mdpi.com/

Sustainability Editorial Office
E-mail: sustainability@mdpi.com
http://www.mdpi.com/journal/sustainability

www.ingramcontent.com/pod-product-compliance
Lightning Source LLC
Chambersburg PA
CBHW040805250926
51850CB00007BA/1083
9783038421115